2nd International Conference on Computational Intelligence & IoT (ICCIIoT 2021)

Journal of Physics: Conference Series Volume 2236

Online
23 – 24 February 2022

ISBN: 978-1-7138-5492-0
ISSN: 1742-6588

Printed from e-media with permission by:

Curran Associates, Inc.
57 Morehouse Lane
Red Hook, NY 12571

Some format issues inherent in the e-media version may also appear in this print version.

Institute of Physics
Dirac House, Temple Back
Bristol BS1 6BE UK

Phone: 44 1 17 929 7481
Fax: 44 1 17 920 0979

techtracking@iop.org

Additional copies of this publication are available from:

Curran Associates, Inc.
57 Morehouse Lane
Red Hook, NY 12571 USA
Phone: 845-758-0400
Fax: 845-758-2633
Email: curran@proceedings.com
Web: www.proceedings.com

TABLE OF CONTENTS

2nd International Conference on Computational Intelligence and Internet of Things (ICCIIoT 2021)

Preface:
We are pleased to present this volume of the selected contributed articles, submitted and presented in the 2nd International Conference on Computational Intelligence and Internet of Things (ICCIIoT 2021) which held on February 23-24, 2022 (initially planned on November 22 & 23, 2021 but postponed due to COVID19) at Department of Electronics and Communication Engineering, North Eastern Regional Institute of Science and Technology (NERIST), Arunachal Pradesh, India, in collaboration with International Association of Academicians (IAASSE), USA. (Conference link: http://icciiot.iaasse.org/.)

The 2nd International Conference on Computational Intelligence and Internet of Things (ICCIIoT 2021) is the 2nd in the series) is a platform for researchers, engineers, academicians as well as industrial professionals from all over the world to present their research results and development activities in Computational Intelligence and Internet of Things. The event (ICCIIoT 2021) took place in hybrid model and thereby allowing some authors to present their papers virtually because of Covid–19 and the government's restriction on any gathering. And by that, organizers fulfilled their commitments with researchers and supports the scientific research. The eminent professors and experts in their fields were invited to deliver their speech at ICCIIoT 2021.

The papers presented in the ICCIIoT 2021 went through strict refereeing and examination resulting in a current acceptance rate of 22.4%. All papers were selected for oral presentation in the conference after an initial review. Every submitted full length paper was sent for peer review to at least two potential reviewers of related areas of expertise once it was passed by the program committee. We are delighted to say that this is in no small part due to the hard work the editorial board and reviewers, in not only refereeing the papers submitted but raising the standard of the quality of papers that are to be published.

This volume of the proceedings from ICCIIoT 2021 will be a very useful contribution for the researches, educators and scientists for enthusing new vistas of research and development in the emerging areas of Computational Intelligence and Internet of Things.

ACKNOWLEDGMENT

The guest editors warmly thank and record their heartfelt deep appreciation to all the eminent guests including the keynote speakers, plenary speakers and other delegates who made their valuable contributions towards the success of the ICCIIoT 2021. It is a great opportunity for us to thank all the reviewers for sparing time from their busy schedule and provide timely and valuable comments to improvise the original versions of the papers towards these extended papers. We would like to thank representatives of IOP also for their collaboration and full cooperation. Last but not the least, as the guest editors of ICCIIoT 2021 proceedings, we are thankful to people who have worked with us in planning and organizing both technical arrangements. We also thank all learned authors for their kind cooperation and contribution. Hope that the quality research work published in this issue will be able to serve something new to Humanity and Science.

Disclaimer:

The authors are solely responsible regarding the authenticity and ethical issues related to their article contained in this volume.

Guest Editors:

Prof. Rajesh Kumar, Department of ECE, NERIST, Arunachal Pradesh, India
Prof. Ashok Ray, Department of ECE, NERIST, Arunachal Pradesh, India
Prof. Nishtha Kesswani, CURAJ, Rajasthan, India
Prof. Frank Lin, CSUSB, USA
Prof. Ashok Patel, Florida Polytechnic University, USA
Prof. Gaurav Trivedi Department of EE, IIT Guwahati, India
Prof. N. Kishorjit Singh Department of CSE, IIIT, Manipur, India

Journal of Physics: Conference Series **2236** (2022) 011002 doi:10.1088/1742-6596/2236/1/011002

PEER REVIEW STATEMNT-ICCIIoT 2021

All papers published in this volume have been reviewed through processes administered by the Editors. Reviews were conducted by expert referees to the professional and scientific standards expected of a proceedings journal published by IOP Publishing.

1. Type of peer review: Double Anonymous
2. Conference submission management system: Morressier
3. Number of submissions received: 58
4. Number of submissions sent for review: 58
5. Number of submissions accepted: 12
6. Acceptance Rate (Submissions Accepted / Submissions Received × 100): 20.7 %
7. Average number of reviews per paper: 2.40
8. Total number of reviewers involved: 18
9. Contact person for queries:
   Name: Rajesh Kumar
   Affiliation: NERIST, India
   Email: icciiot@iaasse.org

Application of Dunkl Wavelets to solve the Integral Equations

M M Dixit, C P Pandey, Abhishek Kumar Pandey and Pranami Phukan

Department of Mathematics
North Eastern Regional Institute of Science and Technology
Nrjuli, Itanagar, Arunachal Pradesh-791109, India

pranamiphukan94@gmail.com

Abstract: The Integral Equation appears in different fields of science and engineering mathematics, for example image processing and the concept of electromagnetic. The technique of wavelet transform to find the solution of integral equations allows creating an algorithm after comparing with known algorithms. This paper is about the uses of the Dunkl Wavelet transforms and associated convolution to establish some suppositions for solution of these equations on distribution spaces. This work defined the method to solve various integral equations by using the Dunkl wavelet transforms.

Keywords: Dunkl transform, Dunkl Wavelet transforms, Dunkl Convolution product.

1. Introduction

The type of equations under Integral sign which contains some unknown functions is known as Integral equations. Generally for the solutions of Integral equations we use Fourier transform, Laplace transform, Hankel transform in distributional sense. From last few years' mathematicians are using wavelets involving some special functions and generalized integral transforms in Pure and Applied Mathematics[2]. In the pure part of Mathematics the concept and formulae of generalized wavelet transforms are involved to explore distribution and function spaces to get the solutions of integral differential equations[1]. To solve some integral equations many authors used classical wavelets, such as by applying Gegenbauer wavelets, Pathak[13] has find the result of Abel integral equation, Mandal[8] studied about various Integral equations and its solutions by making use of fractional calculus and fractional convolution technique in mathematics, also by utilizing Legendre wavelet and related convolution, Pathak et. al.[12] have established the result of generalized Abels integral equation .

To solve definite type of differential equations one can apply Dunkl functions. Pandey et. al.[9] studied about the continuous Dunkl wavelet transform. In quantum physics differential equations are arise and Dunkl wavelet has widely used to solve it. Also Dixit and Pandey[3] studied the generalized Dunkl Wavelet transforms. To see the works mentioned above, by the concept of Dunkl transform and Dunkl wavelet transform, we examine the solution of defined integral equations. We are taking the concept from our work where we i.e. Pandey and Phukan[14] have explained about the formula to solve certain integral equations by utilizing the concept of Bessel wavelet transform and related Hankel transform.

Journal of Physics: Conference Series **2236** (2022) 012001 doi:10.1088/1742-6596/2236/1/012001

2. BACKGROUND

We define the space $L_{p,\alpha} = L_p(d\mu_\alpha),\ \ 1 \le p \le \infty,$ of real measurable functions ϕ on R, where

$$\|\phi\|_{p,\alpha} = \left[\int_R |\phi|^p \, d\mu_\alpha(t_1)\right]^{1/p} < \infty, \quad 1 \le p < \infty \ . \tag{1}$$

and $\|\phi\|_{L_{\infty,\alpha}} = ess \sup_{x \in R} |\phi(t_1)| \ \ if\ p = \infty$.

The Dunkl transform F_α of a function $\phi \in L_{1,\alpha}(R)$, is given by

$$F_\alpha \phi(\lambda) = \hat{\phi}(\lambda) = \int_R E_\alpha(-i\lambda t_1)\phi(t_1) d\mu_\alpha(t_1); \ \ \lambda \in R \ . \tag{2}$$

The inversion for the above equation is given by

$$F_\alpha^{-1}\left(\hat{\phi}(\lambda)\right) = \left(\hat{\phi}(\lambda)\right)^{\vee} = \phi(t_1) = \int_R E_\alpha(i\lambda t_1)\hat{\phi}(\lambda) d\mu_\alpha(\lambda) . \tag{3}$$

For $\phi \in L_{p,\alpha}(R)$ and $\omega \in L_{q,\alpha}(R)$, the Dunkl convolution

$$\phi *_\alpha \omega(t_1) = \int_R \tau_{u_1}\phi(-t_2)\omega(t_2) d\mu_\alpha(t_2) . \tag{4}$$

here τ_{t_1} is the Dunkl translation and is stated as

$$\tau_{t_1}\phi(t_2) = \phi(t_1, t_2) = \int_R \phi(t_3)\chi_\alpha(t_1, t_2, t_3) d\mu_\alpha(t_3) \tag{5}$$

where $\chi_\alpha(t_1,t_2,t_3) = \left(1 - \beta_{t_1,t_2,t_3} + \beta_{t_3,t_1,t_2} + \beta_{t_3,t_2,t_1}\right)\Delta_\alpha(t_1,t_2,t_3)$

and $\beta_{t_1,t_2,t_3} = \begin{cases} \dfrac{t_1^2 + t_2^2 + t_3^2}{2t_1 t_2}, & if\ t_1, t_2 \in R \setminus 0 \\ 0 & , otherwise \end{cases}$

where Δ_α is the Dunkl kernel as

$$\Delta_\alpha(t_1,t_2,t_3) = \begin{cases} d_\alpha \dfrac{\left(\left[\left(|t_1| + |t_2|\right)^2 - t_3^2\right]\left[t_3^2 - \left(|t_1| - |t_2|\right)^2\right]\right)^{\alpha - \frac{1}{2}}}{|t_1 t_2 t_3|^{2\alpha}}, & if\ |t_3| \in A_{t_1 t_2} \\ 0 & , otherwise \end{cases}$$

$$d_\alpha = \frac{(\Gamma(\alpha+1))^2}{\left(2^{\alpha-1}\sqrt{\pi}\Gamma\left(\alpha+\frac{1}{2}\right)\right)},$$

and $A_{t_1,t_2} = \left(\left\||t_1|-|t_2|\right\|,|t_1|+|t_2|\right)$.

Also $\int_R |\chi_\alpha(t_1,t_2,t_3)|\, d\mu_\alpha(t_3) \le 4.$

For all $t_1 \in R$ and $\phi \in L_{p,\alpha}(R),\ p \ge 1$

$$\|\tau_{t_1}\phi\|_{p,\alpha} \le 4\|\phi\|_{p,\alpha}.$$

Assume $p,q,r \in [1,\infty[$ and $\frac{1}{r}=\frac{1}{p}+\frac{1}{q}-1,\ \phi \in L_{p,\alpha}(R)$ and $\omega \in L_{q,\alpha}(R)$. Then convolution $\phi *_\alpha \omega(t_1)$ satisfies the following norm inequality

(i) $\|\phi *_\alpha \omega\|_{r,\alpha} \le 4\|\phi\|_{p,\alpha}\|\omega\|_{q,\alpha}.$

Moreover for all $\phi \in L_{1,\alpha}(R)$ and $\omega \in L_{2,\alpha}(R)$, we have

(ii) $(\phi *_\alpha \omega)^\wedge(\lambda) = \hat{\phi}(\lambda)\hat{\omega}(\lambda).$

For a function $\psi \in L_{p,\alpha}(R)$, define the dilation D_α is given by

$$D_\alpha\psi(t_1) = \psi(\rho_1 t_1),\ \rho_1 \in R.$$

Using translation and dilation, the Dunkl wavelet $\psi_{\rho_2,\rho_1}(t_1)$ is stated as

$$\begin{aligned}\psi_{\rho_2,\rho_1}(t_1) &= \tau_{\rho_2} D_{\rho_1}\psi(t_1) = \tau_{\rho_2}\psi(\rho_1 t_1) \\ &= \int_R \psi(\rho_1 t_3)\chi_a(\rho_2,t_1,t_3)\, d\mu_\alpha(t_3),\ \rho_2 \in R.\end{aligned} \tag{6}$$

For $\psi \in L_{2,\alpha}$ and $C_\psi = \int_{-\infty}^{\infty} \frac{|\hat{\psi}(\lambda)|^2}{|\lambda|} d\mu_\alpha(\lambda) < \infty$. The Parseval's formula for Dunkl wavelet of any $\phi,\omega \in L_{2,\alpha}$, we obtain

$$\int_{-\infty}^{\infty}\int_{-\infty}^{\infty} \left(\tilde{D}_\psi\phi\right)(\rho_2,\rho_1)\overline{\left(\tilde{D}_\psi\omega\right)(\rho_2,\rho_1)}\frac{d\mu_\alpha(\rho_1)d\mu_\alpha(\rho_2)}{\rho_1} = C_\psi\int_{-\infty}^{\infty}\phi(t)\omega(t)\, d\mu_\alpha(t) \tag{7}$$

If $\phi \in L_{2,\alpha}$ then the Inversion formula for Dunkl wavelet transform is

Journal of Physics: Conference Series **2236** (2022) 012001 doi:10.1088/1742-6596/2236/1/012001

$$\phi(u_1) = \frac{1}{C_\psi} \int\int \left(\tilde{D}_\psi \phi\right)(\rho_2, \rho_1) \psi_{\rho_2,\rho_1}(x) \frac{d\mu_\alpha(\rho_1) d\mu_\alpha(\rho_2)}{\rho_1}.$$

where $C_\psi = \int_{-\infty}^{\infty} \frac{|\hat{\psi}(\lambda)|^2}{|\lambda|} d\mu_\alpha(\lambda).$

This section contains the result of definite integral equations by utilizing wavelet transform and its related convolution. And after that we explain the idea on distribution spaces.

3. Application of Dunkl wavelet transforms in Integral Equatons

3.1 Integral equation of Volterra first kind

Assume that

$$\phi(t_1) = \int_0^\infty \tau_y \kappa(t) \omega(t) dt \qquad (8)$$
$$= (\kappa *_\alpha \omega) \ ,$$

here the operator $\tau_y \kappa(t)$ is the operator for translation. By Dunkl wavelet transform of (8) and its related convolution product (7) equation (8) defer

$$\left(\tilde{D}_\psi \phi\right)(\rho_2, \rho_1) = \left(\tilde{D}_\psi \kappa\right)(\rho_2, \rho_1) \left(\tilde{D}_\psi \omega\right)(\rho_2, \rho_1).$$

So

$$\left(\tilde{D}_\psi \omega\right)(\rho_2, \rho_1) = \frac{\left(\tilde{D}_\psi \phi\right)(\rho_2, \rho_1)}{\left(\tilde{D}_\psi \kappa\right)(\rho_2, \rho_1)}. \qquad (9)$$

If we take the inverse Dunkl wavelet transform of (9), we obtained

$$\omega(t) = \tilde{D}_\psi^{-1} \left[\frac{\left(\tilde{D}_\psi \kappa\right)(\rho_2, \rho_1)}{\left(\tilde{D}_\psi \phi\right)(\rho_2, \rho_1)} \right] = \tilde{D}_\psi^{-1} h(\rho_2, \rho_1),$$

where

$$h(\rho_2, \rho_1) = \frac{\left(\tilde{D}_\psi \kappa\right)(\rho_2, \rho_1)}{\left(\tilde{D}_\psi \phi\right)(\rho_2, \rho_1)}.$$

Here taking the inverse of Dunkl wavelet transform

$$\omega(t) = \frac{1}{C_\psi} \int_0^\infty \int_0^\infty h(\rho_2, \rho_1) \psi_{\rho_2,\rho_1}(t) \frac{d\mu_\alpha(\rho_1) d\mu_\alpha(\rho_2)}{\rho_1},$$

where $C_\psi = \int_0^\infty \frac{|\hat{\psi}(\lambda)|^2}{|\lambda|} d\mu_\alpha(\lambda) < \infty$.

Now we can evaluate the RHS of (9) by using the formula stated above.

3.2 Volterra Integral equation of second kind

Assume that

$$\omega(u) = \phi(u) + \int_0^\infty \tau_y \kappa(t)\omega(t)dt, \qquad (10)$$
$$= \phi(u) + (\kappa *_\alpha \omega).$$

Taking the Dunkl wavelet transform of (10) along with making the use of equation (7) we obtained

$$\left(\tilde{D}_\psi \omega\right)(\rho_2,\rho_1) = \left(\tilde{D}_\psi \phi\right)(\rho_2,\rho_1) + \left(\tilde{D}_\psi \kappa\right)(\rho_2,\rho_1)\left(\tilde{D}_\psi \omega\right)(\rho_2,\rho_1)$$

Therefore

$$\left(\tilde{D}_\psi \omega\right)(\rho_2,\rho_1)\left[1-\left(\tilde{D}_\psi \kappa\right)(\rho_2,\rho_1)\right] = \left(\tilde{D}_\psi \phi\right)(\rho_2,\rho_1).$$

Therefore

$$\left(\tilde{D}_\psi \omega\right)(\rho_2,\rho_1) = \frac{\left(\tilde{D}_\psi \phi\right)(\rho_2,\rho_1)}{\left[1-\left(\tilde{D}_\psi \kappa\right)(\rho_2,\rho_1)\right]}. \qquad (11)$$

If we take the inverse of Dunkl wavelet transform of (11), then obtained

$$\omega(t) = \tilde{D}_\psi^{-1}\left[\frac{\left(\tilde{D}_\psi \phi\right)(\rho_2,\rho_1)}{\left[1-\left(\tilde{D}_\psi \kappa\right)(\rho_2,\rho_1)\right]}\right] = \tilde{D}_\psi^{-1}\eta(\rho_2,\rho_1),$$

where

$$\eta(\rho_2,\rho_1) = \frac{\left(\tilde{D}_\psi \phi\right)(\rho_2,\rho_1)}{\left[1-\left(\tilde{D}_\psi \kappa\right)(\rho_2,\rho_1)\right]}.$$

If we use the inverse of Dunkl wavelet transform, we obtained

$$\omega(t) = \frac{1}{C_\psi}\int_0^\infty\int_0^\infty \eta(\rho_2,\rho_1)\psi_{\rho_2,\rho_1}(t)\frac{d\mu_\alpha(\rho_1)d\mu_\alpha(\rho_2)}{\rho_1}.$$

This gives the solution of equation (10).

3.3 Abel's Integral equation

Let

$$\phi(u) = \int_0^u \frac{\omega(u)}{(\tau-u)^\gamma}du = \left(\omega *_\alpha t^{-\gamma}\right),\ 0<\gamma<1, \qquad (12)$$

when $\tilde{D}_\psi\left[t^{-\gamma}\right]$ is given and if we apply the convolution of the Dunkl wavelet transform on (12), such that

$$\left(\tilde{D}_{\psi}\phi\right)(\rho_2,\rho_1)=\left(\tilde{D}_{\psi}\omega\right)(\rho_2,\rho_1)\left(\tilde{D}_{\psi}t^{-\gamma}\right)(\rho_2,\rho_1) \,. \tag{13}$$

Take ψ mother wavelet, then Dunkl wavelet transform of $t^{-\gamma}$ is

$$\left(\tilde{D}_{\psi}t^{-\gamma}\right)(\rho_2,\rho_1)=\varphi(\rho_2,\rho_1) \tag{14}$$

By using (14) in (13), we have

$$\left(\tilde{D}_{\psi}\phi\right)(\rho_2,\rho_1)=\left(\tilde{D}_{\psi}\omega\right)(\rho_2,\rho_1)\varphi(\rho_2,\rho_1) \,.$$

Therefore

$$\left(\tilde{D}_{\psi}\omega\right)(\rho_2,\rho_1)=\frac{\left(\tilde{D}_{\psi}\phi\right)(\rho_2,\rho_1)}{\varphi(\rho_2,\rho_1)} \,.$$

By inverse Dunkl wavelet transform we can evaluate equation (13). So the inverse of Fourier-Jacobi wavelet transform is

$$\omega(t)=\tilde{D}_{\psi}^{-1}\left[\frac{\left(\tilde{D}_{\psi}\phi\right)(\rho_2,\rho_1)}{\varphi(\rho_2,\rho_1)}\right]=\tilde{D}_{\psi}^{-1}\nu(\rho_2,\rho_1),$$

where

$$\nu(\rho_2,\rho_1)=\frac{\left(\tilde{D}_{\psi}\phi\right)(\rho_2,\rho_1)}{\varphi(\rho_2,\rho_1)} \,.$$

If $\phi(t)$ is known, then we can find $\omega(t)$ by inverse wavelet transform.

$$\omega(t)=\frac{1}{C_{\psi}}\int_0^{\infty}\int_0^{\infty}\nu(\rho_2,\rho_1)\psi_{b,a}(t)\frac{d\mu_{\alpha}(\rho_1)d\mu_{\alpha}(\rho_2)}{\rho_1}.$$

On non-identical distribution and function spaces we can compute the result of integral equation only the way, if we explain the generalized convolution and translation operator. Many Mathematicians have studied about it. A, Prasad [11] has worked on the wavelet transform formed by Bessel wavelet as distribution spaces. By operating notion of generalized convolution we can found the result of integral equations.

The convolution $\phi *_{\alpha} \omega$ is defined, if both the functions ϕ and ω are compact supports. For these two generalized functions ϕ and ω, the generalized convolution[10] is in the given form

$$\left(\phi *_{\alpha} \omega\right)(w)=\int_a^b(\tau_w\omega)\omega(\mu)d\mu_{\alpha}.$$

Also for two distributions ϕ and ω the generalized Dunkl transforms of $\phi *_{\alpha} \omega$ is as given below

$$I_{\Phi}\left(\phi *_{\alpha} \omega\right)(t) = (I_{\Phi}\phi)(t)(I_{\Phi}\omega)(t).$$

Furthermore if $\phi(t)$, $\omega(t)$ and $t^{-\gamma}$ are generalized functions, we put the results of integral equations which are acquired earlier by generalized Dunkl wavelet transform, in the generalized functions These are calculated by distributional wavelet transform.

4. Conclusion:

To solve Integral equations we can utilize different wavelet techniques. For the stable and error-free solution of the integral equation, the Dunkl wavelet is the major suitable method. This paper states that by utilizing the continuous Dunkl wavelet transform in the given three integral equations we can get the established results and by that results we can find the error-free solutions of the given three types of Integral equations. The Dunkl wavelet transforms used to solve boundary value problems.

References

[1] Banerji P K and Loonker Deshna, 2005, Distributional Integral transforms, Scientific Publishers, India.

[2] Chui C K, 1992, An Introduction of Wavelets, Academic Press.

[3] Dixit M M and Pandey C P, 2018 Generalized Dunkl Wavelet on N-Dimensional Spaces, *Journal of Computer and Mathematical Science*, 9(8): 941-947.

[4] Estrada R and Kanwal R P, 2000 Singular Integral Equations, Birkhauser, Boston, Basel.

[5] Loonker Deshna and Banerji P K, 2011 On the solution of distributional Abel integral equation by distributional Sumudu transform, *Internat. J. Math. Sci.*, 1-8, Article ID 480528.

[6] Loonker Deshna and Banerji P K, 2014 On the distributional Abel integral equation for distributional Elzaki transform, *J. Indian Math. Soc.,* 81 (1-2), 87-96.

[7] Loonker Deshna and Banerji P K, 2004 Natural transform and solution of integral equations for distribution spaces, *American Journal of Mathematics and Sciences*, 3(1): 5-72.

[8] Mandal N B, Chakrabati A and Mandal N, 1996 A system of generalized Abel's integral equation using fractional calculus, *Appl. Math., Lett.*, 9(5): 1-4.

[9] Pandey C P, Mohan Rakesh and Tripthi Bhairaw Nath, 2014 Continuous Dunkl Wavelet Transform, *International Journal of Current Engineering and Technology*, 4(1): 305-307.

[10] Pathak R S, 2009The Wavelet transform, Atlantis Press, Amsterdam, Paris.

[11] Pathak R S and Pandey C P, 2006 The wavelet transform in a generalized Sobolev space, *J. Indian Math. Soc.*, 73 (3-4): 235-247.

[12] Pathak Ashish, Dixit M M and Upreti P S, 2013 Numerical solution of system of system of generalized Abel's integral equation using Legendre Multiwavelet, *J. Adv. Res. Rese. In Sci. comp.*, 5(1): 28-36.

[13] Pathak Ashish and Singh R K, 2014 Solution of system of generalized Abel's Integral equations using Gegenbauer wavelets, *Internat. J. comp. appl.*, 95(1): 1-4.

[14] Pandey C P, Phukan Pranami and Moungkang Khetjing, 2021 Solution of Integral equations by Bessel wavelet transform, *Advances in Matematics: Scientific Journal*, 10(4): 2245-2253.

Journal of Physics: Conference Series **2236** (2022) 012002 doi:10.1088/1742-6596/2236/1/012002

Retraction

Retraction: Deep Learning-Based Clinical Image Segmentation Methods With Different Modalities (*J. Phys.: Conf. Ser.* **2236** 012002)

Published 11 May 2022

This article has been retracted by IOP Publishing following an allegation that this article may contain tortured phrases [1], masking overlap of other work [2].

IOP Publishing has investigated in line with the COPE guidelines, and agree this article should be retracted.

IOP Publishing wishes to credit the Problematic Paper Screener for bringing the issue to our attention.

The authors agree to this retraction.

[1] Cabanac G, Labbe C, Magazinov A, 2021, arXiv:2107.06751v1

[2] Sumithra M, Malathi S, 2018, A Survey on Medical Image Segmentation Methods with Different Modalitites, IJERT. **6** 2

Retraction published: 11 May 2022

Deep Learning-Based Clinical Image Segmentation Methods With Different Modalities

Dhyanendra Jain[1], Prashant Singh[2] , P.K.Bharti[3]
[1,2]Department of Information Technology
[3]Vice Chancellor, Shri Venkateshwara University
[1,2]Dr. Akhilesh Das Gupta Institute of Technology & Management, New Delhi
[3]Shri Venkateshwara University
dhyanendra.jain@gmail.com[1], prashant.ert@gmail.com[2], vc@svu.edu.in[3]

Abstract – This paper provides a evaluation of the furnished strategies for classifying picture of brain lump the Nuclear-MRI (MRI) and the computed tomography (CT) scanning. The separation of MRI picture of the mind and CT are extensively used as preoperative imaging, in-affected person studies and with inside the use of equipment, within side the area of scientific imaging processing. Separation of MRI and CT picture is a tough venture due to the similarities among the robust and tender institution of muscle anatomy with inside the mind photo. And many components of the identical area within side the image are separated via way of means of the separation of the photo and the direction. Selection of the ideal phase kind constructed at the photo element. This evaluation changed into carried out with the purpose of authorizing a mixture of kind separation of MRI and CT picture. These opinions are classified primarily based totally at the strategies taken into consideration for classification.Imaging of MRI & CT also known as X-ray. The separation of MRI picture of the mind and X-ray are extensively used as preoperative imaging, in-affected person studies and withinside the use of equipment, withinside the area of scientific imaging processing. Separation of MRI and CT picture is a tough venture due to the similarities among the robust and tender institution of muscle anatomy withinside the mind photo. And many components of the identical area withinside the image are separated via way of means of the separation of the photo and the direction. Selection of the ideal phase kind constructed at the photo element. This evaluation changed into carried out with the purpose of authorizing a mixture of kind separation of MRI and CT picture. These opinions are classified primarily based totally at the strategies taken into consideration for classification.

Keywords — DCT, Nuclear-MRI, X-Ray, Positron Emission Tomography, single-photon Emission Computed Tomography , Artificial intelligence, Brain lump.

1.Introduction

These days, the tomography kind like X-Ray, the Nuclear-MRI, the single-photon emission computed tomography & the positron emission tomography, is contributing to straightforward imaging. All kind have specific type of benefits: excessive temporal along with spatial reason with the CT; remarkable tissue[7]

Characterization and non-ionizing transmission with MRI; and massive compassion for molecular imaging with SPECT or PET. Anyway, now no longer even one kind is sufficient to reveal the tough dynamics of human body structure and pathology. Therefore, exceptional forms of hybridization/fusion are vital for better structural, functional, and atomic imaging. An instance of many remedies. These days, tomography modalities along with computed tomography, Nuclear-MRI, unmarried photon emission-computed tomography , and positron emission tomography are used for scientific imaging. All of those sorts have surely identifiable approving circumstances: massive worldly and spatial willpower with *Computed* Tomography; extra special tissue portrayal and nonionizing emission with MRI; and massive affectability for sub-atomic imaging with SPECT or PET. Even one unmarried method is ok to delineate the complicated factors of human body structure and pathology. Henceforth, exceptional forms of hybridization and combination are critical for important auxiliary, utilitarian, and sub-atomic imaging. A case of various healing implants has been offered for X-ray micro modulated luminescence tomography (XMLT). As showed through the achievements of SPECT-CT and PET-CT scanners, the method of mixture imaging is unmistakably compelling and interactive and has had a colossal impact on each exploratory revelation and scientific care. The subsequent 1/2 of-and-1/2 of advances like PET-MRI and XMLT aren't too a ways off, with colossal potential.

Image segmentation is the characterization of an photo through exceptional methods. Numerous analysts have carried out paintings withinside the fields of MRI and CT department the usage of bunching. There are numerous techniques, and it's far apparent that the maximum widespread method is the k-implies bunching calculation. The K- implies grouping computation is an unproven estimate, and it's far used to partition the intrigue variety from the foundation. Before the usage of K- mplies computation, the first step is fractional increasing improvement, that is delivered to the photo to enhance the individual of the photo. The subtractie bunching method is with a purpose to organization the technique that produces the center of the component in mild of the feasible calculation of the records that shows. So a subtractive organization is used to generate the focuses and those focuses are used as part of k-method fractional for the divided photo. At that actual center, the remaining middling channel is joined to the portioned photo to push out any objectionable settings from the photo.

2.Literature Review

Display the principle outline of custom-designed the radiotherapy meant for brain lumps. The calculated version for the glioblastoma improvement became blended with the mathematical characterization of the molecular tolerance version to expose the impact of radiotherapy. Customized the Nuclear-MRI-(MRIs) for a certain individual [1]. It appears on dangers to a solid version, together with the instability with inside the MRI stages. Registered dispersing of plant molecular adhesion potential, blended with a molecular tolerance version, is used to signify drug dosage, which ends up in compliance with Intensity Modulated Radiation Therapy (IMRT). Based on handy scientific information, examine 3 distinct eventualities and layout a custom-designed layout.

Consistent with the 3 hooked-up novel strategies of calculating drug dosage. First, lessen the residual separation of the tumor cells after the mirrored image of the audible tumor molecular thickness. For a while, decrease the tumor cells that separate the predicted survival after light. Third, display the way to regulate the dosage for medicine to decide the proximity of uncovered organs to danger.To our knowledge, that is the primary pastime the use of a private mind tumor demonstration thinking about the instability of tumor improvement and scientific classes in radiotherapy making plans packages right here the use of some differentiating strategies namely, one. The time factor is used for trying out from the returned broadcast the use of the Metropolis-Hasting calculation beginning with the presentation [7], and for the private exercise of plant improvementin [8]. The twin attention method is used for the" Gaussian Prepare Hamiltonian Monte Carlo (GPHMC) calculation". Like such distinction during every repetition, the bad take a look at levels are taken from the preceding P-(Zi). In making plans for Radiotherapy they use sketch out dose strategy, Probabilistic capacity & changed capacity for deciding mobileular density.

Separation taken into consideration the usage of MRI. They did now no longer use a whole lot of strategies. Consideration of the device of category of switch charges can be improved. Consideration ought to receive to whether or not it's far viable for processes to switch extra steady dose together with proton remedy result in IMRT making plans being extra steady with the endorsed dose.

Sérgio Pereira [2], a reproduction separation technique primarily based totally on Convolution Neural Net researches (CNN), assessments three x three small kernels. , showed and facts addition may be very powerful in keeping apart mind tumor in MRI images. Welcome to the Brain lump Segmentation Challenge. Website, which constantly reveals the principle issue of all, institutional, and local improvement within side the Dice resemblance Co-efficient metrics (0.8-0.83- 0.7) with Challenge details set. Into classifying brain lump, have unique mechanisms that make up the mathematical version that isn't a part of the simple facts.

In classifying the mind tumor, counting some species that continuously develops a mathematical version that isn't a part of the simple facts. These fashions regularly upload capability this is steady with the specification and former version. Due to abnormalities, lumps may be categorized as extracellular tissue, subcutaneous and connective tissue [9]. Other strategies depend upon calculation [10][12]. With the supposition that the lumps, atlas ought should be calculated in the course of isolation, because of the formation of component and area of neoplasms [10]-[12]. Implant-improving fashions may be used to calculate their most effect, assisting to become aware of higher atlases [11],[12]. The proximity for the voxels give a standard message for easy segmenting via way of means of Markov Random Fields[9]. extensively utilized Markov Random Fields to distinguish brain lumps later than similarly preliminary photograph category into high-quality voxels, with a history-primarily based totally hobby histogram viable. As cited via way of means of Menze et al. [13], efficient fashions are nicely tailored to unknown facts, however it is able to be hard to indiscriminately translate preceding records into an immediate calculation version [13].

The strategies used right here are mentioned right here. It starts off evolved with a pre-processing section that includes the pressure of the biased area adjustment and making the episode normal. Then, in the course of the calculation, the quantity of schooling tracts is hypocritically prolonged across the schooling leaflets and the usage of the samples for the High Grade Gliomas growth the quantity of non-popular Low Grade Gliomas module. Convolutional Neural Networks is constructed on convolution layers with three x three small kernels to permit for structural depth. That approach, deal with heterogeneity as a result of the visualization of multi-scanner MRI scans the usage of popular stiffness as proven via way of means of Nyul et al. It suggests that that is vital in acquiring the great viable division. Brain lumps they range substantially of their geographical area and characteristic of the building, so it has asked the usage of extra facts to govern such variability.

Draw again is implemented to sturdy and tender muscles. In edema segments the lesion additives are much less focused and feature now no longer progressed the detection via way of means of percent and extensively utilized most effective for MRI photos that don't consist of photos of various techniques. Nicolas Cordier [3], describes a brand new and higher manner to expose computerized separation of brain lumps the use of multi-atlas primarily based totally balloting techniques. In the gathering to keep away from the speculation of the inner seek window, the conservation patch-primarily based totally frame research is utilized in some easy ways: The probabilistic version routinely defines the areas of hobby and provides a bigger extent of the tumor, which adopts an set of rules to achieve the maximum competitive running time though. constrained processing strength and resources.

This approach has been examined in Beta Tumor Image Separation Challenges. High-degree responses are achieved, with a constrained studying section as a consequence proscribing the chance of overcrowding. In addition, the segmentation slide does now no longer consist of any heritage processing. On paper [14] did now no longer recognition on edema additives. That can be performed here, they may do 1) Glioma Segmentation [15] - [17] the use of a gadget gaining knowledge of set of rules skilled offline, 2) Multi-Atlas Segmentation: When utilized in glioma type, gaining knowledge of techniques the gadget faces fundamental factors problem. Initially, quantity of learning facts will be typically less [15]. 2nd, lots of algorithms want a effective offline pc erudition section, that could recognition on overuse. The specific styles of atlases are thrilling as they're capable of address a small education database, and are to be had online, permitting faultless integration of recent instances into the education database. The techniques they use are an powerful manner to routinely divide via way of means of the use of balloting techniques primarily based totally on a couple of atlases.

Journal of Physics: Conference Series **2236** (2022) 012002 doi:10.1088/1742-6596/2236/1/012002

But unsatisfactory this could be due to the fact this approach most effective appears on the distances among patches to make a distinction. This can cause very specific outcomes from any utility that the closest neighbor can use. Bjoern H. Menze [4], has located a probable version for mind lesion in multicolored photos which are typically acquainted with the EM component, a not unusualplace approach of modeling mind photos the use of Gaussian compounds and atlas tissue ability the use of expending-maximization (EM)), to degree a mark describe some other picture. In this version upload a probable atlas for the wholesome tissue through hidden atlas for the wound. Size set of rules with closed shape EM updates statistics. This method gets rid of the vertical atlas of the back and front scales of the wound typically withinside the picture information. It suggests the boundary of the entire damage in every channel, enters the comparative wondering withinside the painful outside display in all ways, an critical element of the collection of a couple of mind tumor photos.
Here additionally advocate discriminatory version extensions to map out the output version to unconventional labels with semantic and organic meanings, along with liquid content material or liquid-stuffed structure, however with out one-to-one interplay withinside the hypo- or hyper-extreme wound regions identified via way of means of the efficient version. An progressed version designed for tumor lesions to higher upload to the facet photos, and an prolonged discriminatory version to be one of the top-notch techniques withinside the BRATS test.
Other techniques had been advanced for not unusual place small and small unfavorable lumps [18] - [21]. Plant differentiation techniques frequently achieve insights into different mind tissues and wound differentiation techniques which have obtained direct precision [22]. Cerebrum sores coming about because of requesting mind issues [23], [24] and stroke [25], [26] are actually similar to glioma sores long and multimodal power styles, but have drawn in little interest to date. Special fashions are probably to come across precisely the adjustments among the advent of the wound and different tissues withinside the facts. Although they require huge quantities of education facts to be able to be powerful in artifact and sturdy segregation and modeling, they had been correctly utilized in tumor differentiation [27] - [31]. The diverse tumor-primarily based totally tumor type techniques use dense, wise voxel capabilities derived from anatomical maps [32] or picture stiffness, along with spatial variations [33], [34] or stiffness profiles, used as inputs in knowledge algorithms and support vector machines [35], slicing bushes consist of [32], [36], [37], or in-intensity gaining knowledge of techniques [38], [39]. The approach used is a brand new calculation approach for the advent of a particular tumor in multi-dimensional photos. The version stocks facts approximately the area of the lesion the use of a particular complete modal from a couple of sizes, i.e., multivariate, the cap potential of wholesome tissue instructions to distinguish first-class tissue withinside the mind. In the increase of tissue type, the version carries a hidden a part of every voxel that carries the identical plant view code into voxel, [40], [41]. With the use of probabilistic version confirms every genetic speculation approximately hyper & hypo-depth wound systems in specific channels. In this manner expands the usual EM type set of rules [42], [43] the use of ability tissue atlas [44] -[46] in instances in which a specific experience of area may be properly defined. A poorly illustrated method to the fragmentation of a dispensed picture via way of means of a plant will increase fashions that enhance neighborhood or brief relationships as in [47], [48]. Repeated tumor increase fashions defined in partial segregation [49] endorse a reserved definition of wound development, and may be used to explain the unfold of particular plant systems in a longitudinal series [50], in addition to the form and importance of boundary plant systems [51]. This may sell a deeper aggregate of designs which are the number one recognition of ailment sequencing and configuration of picture styles

in techniques used to show this kind of process [52].Matthieu- Lê [5], by evaluating the additives of a comments loop version is tough because of the shortage of a version that ought to be enough to acquire correct boundaries, ambiguity of plant phases, and a complicated calculation version, that could appropriately come across complicated plant increase variables. It is aimed toward figuring out the character additives of a magnified plant version, via way of means of taking a pattern of succeeding components additives with the aim of display the Nuclear-MRI of specific human being. opposite possibility ratio is primarily based totally on: 1) the maximum regular overall performance of the distribution reaction calculation the use of the trying out may be explored with inside the area of feasible categories [60]. Hopefully this feature may be used to devise a deliberate remedy. Some latest paintings has been executed on linking tumor improvement fashions to radiation reaction fashions to higher mirror radiation remedy schemes [61] - [63]. Such a method can provide custom remedy designs whilst thinking about the dangers to version parameters. Fuyong Xing [6], a pc-assisted imaging look at of histopathology fashions may also offer first useful resource and progressed diagnosis, for instance, brain lump, pancreatic & neuroendocrine (NET) tumor. Robotized center department is crucial in lots of specific checks such as calculated morphological detail calculations. Even so, proudly owning one remains past the attain of the common person. The novel method is an employer primarily based totally at the look at of stable and prepared fundamental divisions and defenses of standing. Looking on the center picture, we start with profound CNN version to develop the possible frame research, formed via way of means of a repeated local blending system with a view to stand out with inside the statements. Next, the radical segment calculation is hand held to split the man or woman cores becoming a member of the sturdy willpower primarily based totally at the small-scale show and the disabled region version. One of the enormous blessings of the proposed shape is that it's far well suited with loads of colour-coded histopathology picture. Due to CNN's ordinary getting to know factor and the particular pre-display screen mode, the proposed researching machine is huge sufficient to create loads of well-completed scenarios. The method used to check the proposed calculation of the pathology statistics units on a scalable scale the usage of loads of tissue scales and colour schemes, in addition to the in addition behind schedule assessment of human enjoy suggests the unheard of implementation of the proposed approach. Combining the fundamental statistics up and down collectively to reap a key frame research via way of means of thinking about how key constraints stay susceptible or not able to resist them. In addition, the proposed quantity may also face misleading alerts because of specific strengths or the premise of intake on digital-enabled fashions. The small-scale show seemed to be stronger than the preceding PCA-primarily based totally version because of a loss of protest barrier [64], [65]. However, the usage of all of the instruction situations is a waste of small enjoyment on a huge database at some point of operation. KSVD [66] is a famous phrase reference, but it isn't always designed as a discriminatory dictionary-primarily based totally getting to know approach along with enhancing and classification. Here, a analyzing of the radical's analyzing and deep purpose is primarily based totally at the phrases relating to the principle character. Unlike KSVD, this system especially selects the maximum reflective contexts from the configuration database as the premise for reference references. The energy of the phrase reference getting to know approach is executed via way of means of proscribing the blended rectangular mistakes with a small limit. Keeping in thoughts the aim always and efficaciously isolating the specific cores, integrate the preceding version with the first-rate form and the curly version with the gap and the bounds of disgust. A new approach of calculation then again creates a distortion of the form the usage of a totally close-up version that bends in a terrible way, and the form reaches the pinnacle the usage of the form received in advance with inside the miniature show. Core chart go with the drift division chart 1) Sparse Shape-Based Model, 2) Shape Variation, whilst given start-up bureaucracy, the proposed segmentation shape then again creates a outstanding distortion of the shape of a powerful dynamic shape and a shapely shapes. The form grows reliably from the internal cores, one in every context, and progresses to the center ends. In the dynamic shape show, situations pass via way of means of searching on the picture look statistics till we attain a strong country, wherein the feature of the associated power achieves the fundamental recognition; with inside the form acquisition planning, bureaucracy that in addition examine the uncommon country of the previous form bind the situations. This decided on utility of statistics base brushing up and down is efficaciously related to bioclinical picture classification[64], [65].

The drawback of this paper is the belief that each one filopathological picture of slides are in large part huge in size (e.g., ten thousand x ten thousand), with inside the destiny lessen the running time of this computation method the usage of dispensed pc systems. By splitting the complete picture into a couple of blanketed tiles and dispensing them to specific specialists, simultaneous mobileular department may be performed the usage of the ace employee approach within side the pc section dispensed via way of means of Spark [67]. Our destiny undertaking is to put in force the proposed pc-generated method to create loads of gadgets and picture on a huge scale. WVSN, picture are assembled previous for the switch to different hub [68]. Although supply picture are far flung or dispensed to DCT-primarily based totally fashions, the realistic strategies to apply within side the DCT surroundings will lessen the calculation of terrific aspect effects [73]. More recently, techniques had been proposed to combine some picture into the DCT region. T. Jinshan [74] proposed techniques with inside the DCT region however those techniques tolerate positive ugly signs and symptoms along with blurred imaginative and prescient or blocking off residues that contaminate picture quality. The proposed figure [75] known as Max directs to pick out portions with JPEG code incorrectly due to the fact the quantity of notably reputable AC coefficient is invalid rule whilst a huge part of the AC coefficients will be calculated to be 0 at some point of expansion. Alternatively [76], extrade is taken into consideration the usual of separation. In any case, attempt to carry about [77] display that the extrade offers a far worse execution than the numerous torture measures. Here, a general picture integration system is proposed within side the DCT region. Then, the picture barrier with excessive spatial frequencies is fed on via way of means of a distorted picture. The approach of trying out the consistency is taken after growing the character of the crop picture. The check is coming, that is being executed on some formats within side the JPEG design, displaying that our method has accelerated the dimensions of the maximum compact picture.

A new approach of neighborhood duplication of multi-web web page picture integration has been proposed in area of the DCT web web page instead. We are searching on the implementation of the proposed method with specific assessment standards and located that the creation of a aggregate within side the DCT. area is higher than traditional strategies while thinking about DCT and the great school room techniques consist of DWT, SIDWT, and NSCT, in phrases of viewing great and amount parameters. Additionally, the proposed approach is less complicated to create and computer-generated as soon as supply photographs are encoded with JPEG editing, mainly on far off visible sensor systems. Compatible with JPG. In the Shannon-Nyquist's complexity theoretical theory, the super paintings of Donaho and Candes [69], [78], [79] indicates that a complete-fledged seek approach can correctly retrieve the flag irrespective of its fee low. The fulfillment of included acquisition relies upon lying on belief that the flag is

Journal of Physics: Conference Series **2236** (2022) 012002 doi:10.1088/1742-6596/2236/1/012002

not enough (else may need to change minor), & size lattice fill the RIP that the drives show. By proscribing the 11 stage of flag, Now the solution to the problem of flag salute is kind. As a fashionable rule, maximum of the signs and symptoms met the requirement for complete discovery, and significant use of the mixed experimental speculation has been demonstrated [80]. The key to the mixed acquisition is the ordinary discount of l1. The l1-standard, additionally known as Manhattan eliminate, calculates the sum of the vertical segments of the vector x. Undoubtedly, both `` small problems " are used withinside the above method. However, the approach isn't always as perplexing because it seems. The sub-trouble are quadratic type, and it has the closed frame research with a few low predictability, even as the 2nd sub-trouble might be resolved the usage of the alleged discount function [81].compare it among CT & MRI it will be tender tissue & non-ionizing one. In any of the cases, MRI showing time will be higher than that of CT will give that , X-ray examining utilizes RF heartbeat to make X-ray signals, and confirmation of RF energy is impeded through X-ray real science. This is consistent with bendy X-ray execution. On the elective hand, CT is speedier. Current CT gets bunches of hypotheses in sync with second, and prevalent CT relaxation realities make the photograph versatile from a fragmented data set [85].In the occasion that MRI is mixed with CT, we are hoping that the CT and MRI capabilities may be seamlessly linked. In this observe, we added the CT-MRI imaging technique. It makes use of an photo from a unmarried factor of view as a priori facts, it is able to examine every other photo relying at the drawing close penalty. To summarize, facts misplaced in a single manner of questioning may be properly repaired with a priori photo from every other photo, in keeping with those strains frees the want for every approach primarily based totally on a simple size of a given photo great. This observe guarantees to lessen radiation publicity and quicker MRI directly.While we've currently studied the duplicate of CT-MRI compound imagery, the maximum formidable awareness is on omnitomography [84], that is an exceptional aggregate of X-ray, N-MRI, PET, SPECT, US, Optical-Imaging & the sky's the limit from there[2]. Each recording method is simply one a part of the complete frame research. In an included shape, statistics is gathered simultaneously, shared throughout all embedded channels, generating extra beneficial synergistic, simple, mobileular characteristic and atomic shape facts. Step-through-step commands for combining near institutions among photographs received in diverse approaches and the way to make complete use of all preceding facts may be an exciting subject matter for our destiny research. In the planned approach of removal of brain lump consists of related processes, preparation, stepped forward fluid glide amendment, morphological sports and post-implant mind transplant confirmation[70]. So we speak the aforementioned schemes in detail: Watershed Transformers is a totally powerful approach of category that strategies the sector of numerical morphology. The smart concept of this modification could be very straightforward: while we all assume an image as a scene or a geomorphometry. aid, while the panorama of every location is correctly diagnosed with its darker state, and recollect the gradual rainfall within side the location, then the wetlands are specific strains The watershed shift is guaranteed a practical photo revision, with the aim that the rims of the suction vessel be located with inside the awareness location of the pinnacle slope. This alternate has been completely applied as a part of quite a few imaging modalities, such as the separation of scientific imaging taken from Nuclear-MRI, for the reason that various positive situations it have: extremely simple, natural, fast, uniform and powerful. Complete separation of the photo in remoted areas irrespective of the terrible difference, due to preserving the strategic distance from the prerequisite of some sort of joining. Another critical degradation associated with watershed transformation is immoderate fragmentation and negative area of tremendous regions with low

segmentation frequencies that frequently gift photographs of the MR cerebrum. The novel approach is examined on a wonderful internet site with photographs of one hundred twenty boils. The plant elements of the MR photo are transparent, performing as white shading. This element has a far better energy than the specific areas of the photo. The calculations right here can efficaciously eliminate mind tumor with 98 percent curacy in specific age groups.
This paper confirmed a alternate in fluid adjustments to eliminate the brain lump of MR photographs in mild of separation & morphological director. Further, contrast amendment is set, this software can differentiate brain lumps from MR photographs into specific age groups. The proposed approach presents extra facts on mind tumor and assists professionals in determining, remedy shape and standing of tumor monitoring. The check comes to reveal that the proposed method can offer very last, easy and correct very last research with a low calculation problem. The proposed challenge may be limitless through acquiring partition facts to healthy the cut up photo retrieval in a specific state. Increasing numbers of dynamic healing photographs In Swathi et. Al, [71] the complete Brain Tumor Detection and Distribution Strategy Using Histogram Thresholding and Artificial Neural Net research technique. In this Paper, dependent frame research is a vital a part of the Artificial Neural Net research. The extrade is primarily based totally at the ordinary use of precise areas (ROIs) in the tumor vicinity in MRI snap shots the usage of the Histogram Thresholding technique. Create all of the rous, positioned in an excavated vicinity containing a drawing of the plant and excellent team spirit is extracted and turn out to be the center. All rous then a group of PDF abrasions of every cerebrum plant within side the MR photo. These weights are worn as a version machine for switching the Nutrition Net research. Accessibility paintings are primarily based totally on Pre- frequency distribution graph image segmentation & ANN research of the mind photo class, brain lump approval.

3.PROPOSED TECHNIQUE

In this paper, we propose research techniques for mind tumor detection, ID and order. The number one technique relies upon at the incorporated machine of photo settlement statistics, at the same time as the opposite relies upon at the reconstruction of the evolved neural machine evolved.In fragmentation may be visible as a method of subdividing an photo into some key additives to facilitate essential trying out [86]. There are about seven instructions to be had for segmentation techniques, consisting of possibilities for pictureel trademark techniques [87], spatial associated techniques, series techniques, speculation-primarily based totally techniques, stage putting hybrid techniques [90]. Over time, from an individual's factor of view, a person's view of the content material of the photo is observed with the aid of using a excellent readiness of the area closest to it's indoors systems as opposed to great pictureel snap shots and their blurry snap shots. i.e. is a primary machine for outlining a subdivision technique which could correctly use each neighborhood spatial statistics and pictureel popularity statistics.

Image popularity control is a puzzling hassle that isn't described in a selected manner or one preceding approach. For the usage of each dimensions and neighborhood pictureel structure statistics, this text shows a approach for classifying snap shots with the aid of using searching at a aggregate of vector amount and aspect acquisition. The method captures small limitations separated with the aid of using a photo inclusive of coping with articles, and makes use of each format statistics and dim esteem statistics to assemble decrease rectangular detail vectors. Starting with the bat, the phase vectors are divided into models, in particular the threshold layout and the non-aspect layout, with the very last identity frame. Alternatively, Boundary Detection (BD) calculations are generated to be subdivided to the quantity that the sub-vector curves of the

Journal of Physics: Conference Series **2236** (2022) 012002 doi:10.1088/1742-6596/2236/1/012002

threshold layout and the Vector Quantization (VQ) technique are proven to split the sub-vector layout sub-piece. Finally, the break up end result is knowledgeable with the aid of using a aggregate of VQ effects statistics and BD comes. The advent of VQ at first advanced from facts presses and laptop exchanges [95, 96]. In this paper, the status quo of VQ is normal of gratifying a layout organization challenge with the aid of using figuring out vectors of the decrease square. The artical advised a system-on-module sensor machine to recognize VQ calculation flexibly. Edge popularity is a not unusual place and pressing indicator of the photo method. On paper, we installation an aspect popularity frameresearch to differentiate the threshold layout and non-aspect layout vectors, and in addition awareness the curved aspect on the threshold vectors with the aid of using the BD figure. From one factor of view, the end result of this phase offers extra boundary statistics without the impact of fragmentation. Then again, the aggregate of statistics among the BD impact and the VQ impact guarantees a greater encouraging break up impact. Reproduction trying out and associated reflections are directed at programs with inside the control of healing snap shots within side the trying out authorization phase. Image remedy amendment is taken into consideration a prerequisite for diagnostic photo recuperation and different healing programs [97, 98]. Last Within side the mind the final purpose of comparing the proposed differentiation approach, several paper-primarily based totally exams and programs to distinguish MRI snap shots of the human mind, consisting of recreated healing snap shots from Brain Web and some actual mind MRI snap shots. By trying out the feasibility is likewise feasible and associated with the FLICM approach, the legitimacy & positive situations of projected approach are confirmed. In this experiment, we evolved a way of classifying snap shots with the aid of using thinking about a composite shape, which changed into supposed to mix the consequences of VQ & BD to acquire the separation function. Produced method changed into authorized with inside the above research, which covered probes that mimic healing snap shots and analyzes of actual recovery snap shots. Additionally, essential photo restrict statistics changed into now no longer to be had for the merchandising of clinical programs. In short, the proposed approach has been correctly close right all the way down to control each re-imagined retrieval snap shots and actual clinical snap shots, and the extra photo restrict statistics changed into closed promising extra photo processing.

This paper indicates the class machine whilst thinking about the aggregate of vector quantization and aspect popularity. There are key factors of the proposed approach: because the maximum essential issue, the department impact is removed with the aid of using the combination of the statistics to well known the impact of the essential vicinity shape of the small fragments and the essential research end result of gritty inner statistics; Second of all, moreover, in compiling evaluation questions, the end result of the proposed strategic department offers photo-restrict statistics for added programs which might be much more likely to be in particular not unusual place in traditional techniques. The approval of this approach is validated within side the experimental vicinity to be used in healing snap shots and is related to the FLICM method. Completion of the check occurs, the proposed technique exceeds the FLICM approach with the aid of using 3% TPF and 4 % TNF is a picture graph-sharing application, with extra photo restrict statistics acquired for photo processing. The photo class approach we've created on this paper need to be useful within side the photo method, mainly with inside the photo recovery method. To enhance the presentation of this software are clinical photo that put together for picture graph registration and entertainment.

4.CONCLUSION

This section present the effort made by other investigator associated to segmentation of clinical images. At this time of all referenced documents is being summarized. It has been present a lot of type of segmentation by use of the different method. to conclude, it present one of the new idea for various portion of division. In view of the fact that the innovation falling in it's predominant manner, however a little slacking in the tracking down the specific outcomes in division. In our method endeavor to beat few downsides in regards to precision and using time productively of the framework. There are numerous strategies present to formulate it unextracted by foes, however it requires numerous fresh calculations to generate precise division. The future strategy give additional a focus on issues to settle. The future calculation performs better compared to presented division calculations.

In our artical we din't accomplished a partial outline of various divisions for X-ray cerebrum picture with test exhibited the various techniques which can section the growth picture adequately and give accurate result. In our method loosened up for new computation for cerebrum growth division which will give better capable result than that of current methods in not really far off future. In our method show's Execution time will be well thought-out to take a gander at this framework capably. As the determination cancer is a confounded and delicate task, precision and steadfastness are continually given out a lot of importance. Consequently a complex technique that high lights for developing numerous overwhelming picture division framework is for the most part searched for.

REFERENCES

[1] Matthieu Lê et al., —Personalized Radiotherapy Planning Based on a Computational Tumor Growth Model‖, in IEEE Transactions On Medical
[2] Sérgio Pereira et al., Brain Tumor Segmentation Using Convolutional Neural Netresearchs in MRI Images, IEEE Transactions On ClinicalImaging, VOL. 35, NO. 5, 2016
[3] Nicolas Cordier et al., —A Patch-Based Approach for the Segmentation of Pathologies: Application to Glioma Labelling‖, IEEE Transactions
[4] Bjoern H. Menze et al., —A Generative Probabilistic Model and DiscriminativeExtensions for Brain Lesion Segmentation—With Application to Tumor and Stroke‖ IEEE Transactions On ClinicalImaging, VOL. 35, NO. 4, 2016.
[5] Matthieu Lê et al., —MRI Based Bayesian Personalization of a Tumor Growth Model‖, IEEE Transactions On ClinicalImaging, VOL. 35, NO. 10, 2016.
[6] Fuyong Xing et al., —An Automatic Learning-Based Frameresearch for Robust Nucleus Segmentation‖, IEEE Transactions On ClinicalImaging, VOL. 35, NO. 2, 2016
[7] C. E. Rasmussen, —Gaussian processes to speed up hybrid Monte Carlo for expensive Bayesian integrals,‖ Bayesian Statist., vol. 7, pp. 651–659, 2003.
[8] M. Lê et al., —Bayesian personalization of brain tumor growth model,‖ in Proc. Int. Conf. Med. Image Comput. Comput. Assist. Intervent., 2015, pp. 424–432.

[9] M. Prastawa *et al.*, —A brain tumor segmentation frameresearch based on outlier detection,‖ *Med. Image Anal.*, vol. 8, no. 3, pp. 275–283, 2004.
[10] B. H. Menze *et al.*, —A generative model for brain tumor segmentation in multi-modal images,‖ in *ClinicalImage Computing and Comput.-Assisted Intervention-MICCAI 2010*. New York: Springer, 2010, pp. 151–159.
[11] A. Gooya *et al.*, —GLISTR: Glioma image segmentation and registration,‖ *IEEE Trans. Med. Imag.*, vol. 31, no. 10, pp. 1941–1954, Oct. 2012.
[12] D. Kwon *et al.*, —Combining generative models for multifocal glioma segmentation and registration,‖ in *ClinicalImage Computing and Comput.-Assisted Intervention-MICCAI 2014*. New York: Springer, 2014, pp. 763–770.
[13] B. Menze *et al.*, —The multimodal brain tumor image segmentation benchmark (BRATS),‖ *IEEE Trans. Med. Imag.*, vol. 34, no. 10, pp.1993–2024, Oct. 2015.
[14] P. Y. Wen *et al.*, —Updated response assessment criteria for high- grade gliomas: Response assessment in neuro-oncology researching group,‖ *J. Clin. Oncol.*, vol. 28, no. 11, pp. 1963–1972, 2010.
[15] B. Menze *et al.*, —The multimodal brain tumor image segmentation benchmark (BraTS),‖ *IEEE Trans. Med. Imag.*, vol. 34, no. 10, pp. 1993–
[16] D. Zikic *et al.*, —Decision forests for tissue-specific segmentation of high-grade gliomas in multi-channel MR,‖ in *MICCAI 2012*. New York: Springer, 2012, pp. 369–376.
[17] N. J. Tustison *et al.*, —Optimal symmetric multimodal templates and concatenated random forests for supervised brain tumor segmentation (simplified) with ANTsR,‖ *Neuroinformatics*, pp. 1–17, 2014.
[18] M. Kaus *et al.*, —Segmentation of meningiomas and low grade gliomas in MRI,‖ in *Proc. MICCAI*, 1999, pp. 1–10.
[19] Y.-F. Tsai, I.-J. Chiang, Y.-C. Lee, C.-C. Liao, and K.-L. Wang, —Automatic MRI meningioma segmentation using estimation maximization,‖ *Proc. IEEE Eng. Med. Biol. Soc.*, vol. 3, pp. 3074–3077, 2005.
[20] E. Konukoglu *et al.*, —Monitoring slowly evolving lumps,‖ *Proc. ISBI*, pp. 1–4, 2008.
[21] B. Bach Cuadra, C. Pollo, A. O. Bardera, O. Cuisenaire, and J. P. Thiran, —Atlas-based segmentation of pathological brain MR images using a model of lesion growth,‖ *IEEE Trans. Med. Imag.*, vol. 23, no. 10, pp. 1301–14, Oct. 2004.
[22] M. Styner and , —3D segmentation in the clinic: A grand challenge II: MS lesion segmentation,‖ *MIDAS J.*, pp. 1–5, 2008.
[23] A. Irimia *et al.*, —Comparison of acute and chronic traumatic brain injury using semi-automatic multimodal segmentation of MR volumes,‖ *J. Neurotrauma*, vol. 28, pp.2287–2306, 2011.
[24] M. Shenton *et al.*, —A review of Nuclear-MRI and diffusion tensor imaging findings in mild traumatic brain injury,‖ *BrainImag. Behav.*, vol. 6, pp. 137–192, 2012.
[25] T. D. Farr and S. Wegener, —Use of Nuclear-MRI to predict outcome after stroke: A review of experimental and clinicalevidence,‖ *J. Cerebr Blood Flow Metab.*, vol. 30, pp. 703–717, 2010.
[26] I. Rekik, S. Allassonnière, T. K. Carpenter, and J. M. Wardlaw, —Clinicalimage analysis methods in MR/CT-imaged acute-subacute ischemic stroke lesion: Segmentation, prediction and insights into dynamic evolution simulation models. A critical appraisal,‖ *NeuroImage: Clinical*, 2012.

[27] D. Cobzas, N. Birkbeck, M. Schmidt, M. Jagersand, and A. Murtha, —3D variational brain tumor segmentation using a high dimensional feature set,‖ in *Proc. ICCV*, 2007, pp. 1–8.

[28] A. Lefohn, J. Cates, and R. Whitaker, —Interactive, GPU-based level sets for 3D brain tumor segmentation,‖ in *Proc. MICCAI*, 2003, pp. 564–572.

[29] L. Gorlitz, B. H. Menze, M.-A. Weber, B. M. Kelm, and F. A. Hamprecht, —Semi-supervised tumor detection in magnetic resonance spectroscopic images using discriminative random fields,‖ in *Proc DAGM.* New York: Springer, 2007, LNCS, pp. 224–233.

[30] C. Lee, S. Wang, A. Murtha, and R. Greiner, —Segmenting brain lumps using pseudo conditional random fields,‖ in *Proc MICCAI.* New York: Springer, 2008, vol. 5242, LNCS, pp. 359–366.

[31] M. Wels, G. Carneiro, A. Aplas, M. Huber, J. Hornegger, and D. Comaniciu, —A discriminative model-constrained graph cuts approach to fully automated pediatric brain tumor segmentation in 3D MRI,‖ in *Proc MICCAI.* New York: Springer, vol. 5241, LNCS, pp. 67–75.

[32] D. Zikic, B. Glocker, E. Konukoglu, A. Criminisi, C. Demiralp, J. Shotton, O. M. Thomas, T. Das, R. Jena, and S. J. Park, —Decision forests for tissue-specific segmentation of high-grade gliomas in multichannel MR,‖ in *Proc. MICCAI*, 2012.

[33] E. Geremia *et al.*, —Spatial decision forests for MS lesion segmentation in multi-channel magnetic resonance images,‖ *Neuroimage*, vol. 57, pp. 378–90, 2011.

[34] E. Geremia, B. H. Menze, and N. Ayache, —Spatially adaptive random forests,‖ in *Proc. 2013 IEEE 10th Int. Symp. Biomed. Imag.*, 2013, pp.1344–1347.

[35] S. Bauer *et al.*, —Multi-scale modeling for image analysis of brain tumor studies,‖ *IEEE Trans. Bio-Med. Eng.*, vol. 59, no. 1, pp. 25–29, Jan. 2012.

[36] W. Wu, A. Y. Chen, L. Zhao, and J. J. Corso, —Brain tumor detection and segmentation in a conditional random fields frameresearch with pictureelpairwise affinity and superpictureel-level features,‖ *Int. J. Comput. Assist. Radiol. Surg.*, pp. 1–13, 2013.

[37] N. J. Tustison *et al.*, —Optimal symmetric multimodal templates and concatenated random forests for supervised brain tumor segmentation (simplified) with ANTSR,‖ *Neuroinformatics* vol. 13, no. 2, pp. 209– 225, Apr. 2015. P. Dvorak and B. H. Menze, —Structured prediction with convolutional neural netresearchs for multimodal brain tumor segmentation,‖ in *Proc MICCAI Med. Comput. Vis. Researchshop*, 2015.

[38] G. Urban, M. Bendszus, F. A. Hamprecht, and J. Kleesiek, —Multimodal brain tumor segmentation using deep convolutional neural netresearchs,‖ in *Proc. MICCAI BRATS*, 2014, pp. 31–35.

[39] T. Riklin-Raviv *et al.*, —Joint segmentation via patient-specific latent anatomy model,‖ in *Proc MICCAI-PMMIA*, 2009, pp. 244–255.

[40] T. Riklin-Raviv, K. Van Leemput, B. H. Menze, W. M. Wells, and P. Golland, —Segmentation of image ensembles via latent atlases,‖

[41] W. Wells, W. Grimson, R. Kikinis, and F. Jolesz, —Adaptive segmentation of MRI data,‖ in *Proc Comput. Vis. Virtual Reality Robot. Med.*, 1995, pp. 57–69.

[42] W. M. Wells, W. E. L. Grimson, R. Kikinis, and F. A. Jolesz, —Adaptive segmentation of MRI data,‖ *IEEE Trans.Med. Imag.*, vol. 15, no. 4, pp. 429–442, 1996.

[43] J. Ashburner and K. Friston, —Multimodal image coregistration and partitioning—A unified frameresearch,‖ *Neuroimage* vol. 6, no. 3, pp. 209–217, Oct. 1997.

[44] [45] K. Van Leemput, F. Maes, D. Vandermeulen, and P. Suetens, —Automated model-based bias field correction of MR images of the brain,‖ *IEEE Trans. Med. Imag.*, vol. 18, no. 10, pp. 885–896, Oct. 1999.

[45] K. M. Pohl, J. Fisher, W. Grimson, R. Kikinis, and W. Wells, —A Bayesian model for joint segmentation and registration,‖ *Neuroimage*, vol. 31, pp. 228–239, 2006.

[46] Y. Tarabalka, G. Charpiat, L. Brucker, and B. H. Menze, —Spatio- temporal video segmentation with shape growth or shrinkage constraint,‖

[47] E. Alberts, G. Charpiat, Y. Tarabalka, M. A. Weber, C. Zimmer, and M.B. H. Youdim, —A nonparametric growth model for estimating tumor growth in longitudinal image sequences,‖ in *Proc MICCAI Brain Lesions Researchshop (BrainLes)*. New York: Springer, 2015, LNCS.

[48] B. H. Menze, E. Stretton, E. Konukoglu, and N. Ayache, —Image- based modeling of tumor growth in patients with glioma,‖ in *Optimal Control in Image Processing*, C. S. Garbe, R. Rannacher, U. Platt, and T. Wagner, Eds. Heidelberg, Germany: Springer, 2011.

[49] E. Konukoglu *et al.*, —Image guided personalization of reaction- diffusion type tumor growth models using modified anisotropic eikonal equations,‖ *IEEE Trans. Med. Imag.*, vol. 29, no. 1, pp. 77–95, Jan. 2010.

[50] A. Gooya *et al.*, —GLISTR: Glioma image segmentation and registration,‖ *IEEE Trans. Med. Imag.*, vol. 31, no. 10, pp. 1941–1954, Oct.2012.

[51] B. H. Menze *et al.*, —A generative approach for image-based modeling of tumor growth,‖ in *Proc. IPMI*, 2011.

[52] E. Konukoglu et al., —Efficient probabilistic model personalization integrating uncertainty on data and parameters: Application to eikonaldiffusion models in cardiac electrophysiology,‖ Progr. Biophys. Molecular Biol., vol. 107, no. 1, pp. 134–146, 2011.

[53] D. Neumann et al., —Robust image-based estimation of cardiac tissue parameters and their uncertainty from noisy data,‖ in MICCAI. New York: Springer, 2014, pp. 9–16.

[54] B. H. Menze, K. Van Leemput, A. Honkela, E. Konukoglu, M.-A. Weber, N. Ayache, and P. Golland, —A generative approach for image-based modeling of tumor growth,‖ in IPMI. New York: Springer, 2011, pp. 735–747.

[55] H. L. Harpold, E. C. Alvord Jr, and K. R. Swanson, —The evolution of mathematical modeling of glioma proliferation and invasion,‖ J.Neuropathol. Exp. Neurol., vol. 66, no. 1, 2007.

[56] D. Corwin et al., —Toward patient-specific, biologically optimized radiation therapy plans for the treatment of glioblastoma,‖ PloS One, vol.8, no. 11, p. P. e79115, 2013.

[57] E. Konukoglu et al., —Image guided personalization of reaction- diffusion type tumor growth models using modified anisotropic eikonal equations,‖ IEEE Trans. Med. Imag., vol. 29, no. 1, pp. 77–95, Jan. 2010.

[58] M. Lê, H. Delingette, J. Kalpathy-Cramer, E. Gerstner, T. Batchelor, J. Unkelbach, and N. Ayache, —Bayesian personalization of brain tumor growth model,‖ in MICCAI. New York: Springer, 2015.

[59] M. Lê, J. Unkelbach, N. Ayache, and H. Delingette, —GPSSI: Gaussian process for

sampling segmentations of images,‖ in MICCAI. New York: Springer, 2015.

[60] J. Unkelbach et al., —Radiotherapy planning for glioblastoma based on a tumor growth model: Improving target volume delineation,‖ Phys.Med. Biol., vol. 59, no. 3, p. 747, 2014.

[61] [62] J. Unkelbach et al., —Radiotherapy planning for glioblastoma based on a tumor growth model: Implications for spatial dose redistribution,‖ Phys. Med. Biol., vol. 59, no. 3, p. 771, 2014.

[62] [63] R. Rockne, E. AlvordJr., J. Rockhill, and K. Swanson, —A mathematical model for brain tumor response to radiation therapy,‖ J. Math. Biol., vol. 58, no. 4-5, pp. 561–578, 2009.

[63] S. Zhang, Y. Zhan, M. Dewan, J. Huang, D. N. Metaxas, and X. S. Zhou, —Deformable segmentation via sparse shape representation,‖ in *Int. Conf. Med. Image Comput. Comput. Assist. Intervent. (MICCAI)*, 2011, pp. 451–458.

[64] S. Zhang, Y. Zhan, and D. N. Metaxas, —Deformable segmentation via sparse shape representation and dictionary learning,‖ *Med. Image Anal.*, vol. 16, no. 7, pp. 1385–1396, 2012.

[65] M. Aharon, M. Elad, and A. Bruckstein, —K-SVD: An algorithm for designing overcomplete dictionaries for sparse representation,‖ *IEEE Trans. Signal Process.*, vol. 54, no. 11, pp. 4311–4322, Nov. 2006.

[66] M. Sapkota, F. Xing, F. Liu, and L. Yang, —Skeletal muscle cell segmentation using distributed convolutional neural netresearch,‖ in *High Performance Comput. (HPC) Researchshop Int. Conf. Med. Image Comput. Comput. Assist. Intervent. (MICCAI)*, 2015.

[67] Liu Cao, Longxu Jin, Hongjiang Tao, Guoning Li, Zhuang Zhuang, and Yanfu Zhang, —Multi-Focus Image Fusion Based on Spatial Frequency in Discrete Cosine Transform Domain,‖ IEEE SIGNAL PROCESSING LETTERS, VOL. 22, NO. 2, FEBRUARY 2015.

[68] Yang Lu, Jun Zhao And Ge Wang, —Edge-Guided Dual-Modality Image Reconstruction‖, Received October 27, 2014, accepted November 3, 2014, date of publication November 20, 2014, date of current version November 26, 2014.

[69] J.Mehena and M. C. Adhikary,‖ Brain Tumor Segmentation and Extraction of MR Images Based on Improved Watershed Transform‖, IOSR

[70] Journal of Computer Engineering (IOSR-JCE) e-ISSN: 2278-0661,p- ISSN: 2278-8727, Volume 17, Issue 1, Ver. 2 (Jan – Feb. 2015), PP 01-05.

[71] Swathi P S, Deepa Devassy, Vince Paul and Sankaranarayanan P N, —Brain Tumor Detection and Classification Using Histogam Thresholding and ANN‖ (IJCSIT) International Journal of Computer Science and Information Technologies, Vol. 6 (1) , 2015, 173-176.

[72] Ailing De and Chengan Guo, —An image segmentation method based on the fusion of vector quantization and edge detection with applications to clinicalimage processing‖, Int. J. Mach. Learn. & Cyber. (2014) 5:543–551.

[73] M. B. A. Haghighat, A. Aghagolzadeh, and H. Seyedarabi, —Multi- focus image fusion for visual sensor netresearchs in DCT domain,‖ Comput.

[74] T. Jinshan, —A contrast based image fusion technique in the DCT domain,‖ Digital Signal Process., vol. 14.3, pp. 218–226, 2004.

[75] Y. Phamila and R. Amutha, —Discrete cosine transform based fusion of multi-focus images for visual sensor netresearchs,‖ Signal Process., pp. 161–170, 1995.

[76] M. B. A. Haghighat, A. Aghagolzadeh, and H. Seyedarabi, —Real- time fusion of multi-focus images for visual sensor netresearchs,‖ in 6th IEEE Iranian Machine Vision and Image Processing (MVIP), 2010.

[77] W. Huang and Z. Jing, —Evaluation of focus measures in multi-focus image fusion,‖ Patt. Recognit. Lett., vol. 28.4, pp. 493–500, 2007.

[78] D. L. Donoho, ``Compressed sensing," IEEE Trans. Inf. Theory, vol. 52, no. 4, pp. 1289_1306, Apr. 2006.

[79] 3E. J. Candes, J. Romberg, and T. Tao, ``Robust uncertainty principles: Exact signal reconstruction from highly incomplete frequency informa- tion," IEEE Trans. Inf. Theory, vol. 52, no. 2, pp. 489_509, Feb. 2006.

[80] Y. C. Eldar and G. Kutyniok, Compressed Sensing: Theory and Applica- tions. Cambridge, U.K.: Cambridge Univ. Press, 2012.

[81] W. Yin, S. Osher, D. Goldfarb, and J. Darbon, ``Bregman iterative algorithms for `1-minimization with applications to compressed sensing," SIAM J. Imag. Sci., vol. 1, no. 1, pp. 143_168, 2008

[82] W. Guo and W. Yin, ``Edge guided reconstruction for compressive imaging," SIAM J. Imag. Sci., vol. 5, no. 3, pp. 809_834, Jul. 2012.

[83] J. Canny, ``A computational approach to edge detection," IEEE Trans. Pattern Anal. Mach. Intell., vol. PAMI-8, no. 6, pp. 679_698, Nov. 1986.

[84] G. Wang et al., ``Towards omni-tomography_Grand fusion of multiple modalities for simultaneous interior tomography," PLoS ONE, vol. 7, no. 6, p. e39700, 2012.

[85] H. Yu and G. Wang, ``Compressed sensing based interior tomography," Phys. Med. Biol., vol. 54, no. 9, p. 2791, 2009.

[86] Zhang YJ (2001) Image segmentation. Science Press, Beijing, pp 6–58

[87] Gao XL, Wang ZL, Liu JW (2011) Algorithm for image segmentation using statistical models based on intensity features. Acta

[88] Zhao J, Shao FQ, Zhang XD (2011) Vector-valued images segmentation based on improved variational GAC model. Control Decision 26(6):909–915

[89] Zhao J, Shao FQ, Zhang XD (2011) A fast C-V model for vectorvalued image segmentation. J Comput Aided Design Comput Gr23(11):1869–1874

[90] Wu Y, Xiao P, Wang CM (2010) Segmentation algorithm for SAR images based on the persistence and clustering in the contourlet domain. Acta Optica Sinica 30(7):1977–1983

[91] Thornton J, Srbic A (2013) Spatial pooling for greyscale images. Int J Mach Learn Cybern 4(3):207–216

[92] Gamio J, Belongie S,Majumdar S (2004) Normalized cuts in 3D for spinal MRI segmentation. IEEE Trans Med Imaging 23(1):36–44

[93] Veksler O (2000) Image segmentation by nested cuts. In: Proceedings of IEEE Conference on Computer Vision and Pattern Recognition. IEEE Press, Hoboken, pp 339–344

[94] Wang S, Lu HH, Yang F (2011) Superpictureel tracking. In: Proceedings of IEEE International Conference on Computer Vision. IEEE Press, Hoboken, pp 1323–1330

Journal of Physics: Conference Series **2236** (2022) 012002 doi:10.1088/1742-6596/2236/1/012002

[95] Chen SX (2009). Vector quantization and its application to image processing. University of Electronic Science and Technology, Chengdu

[96] Gray RM (1984) Vector quantization. ASSP Mag IEEE 1(2):4–29

[97] Luo SQ, Zhou GH (2003) Clinicalimage processing and analysis. Science Press, Beijing, pp 67–68

[98] Zhang Q, Lu ZT, Chen C (2011) Spinal MRI segmentation based on local neighborhood information and gaussian weighted Chi squaredistance. Chin J Biomed Eng 30:358–362

Speech Emotion Based Sentiment Recognition using Deep Neural Networks

Ravi Raj Choudhary, Gaurav Meena, and Krishna Kumar Mohbey

Department of Computer Science, Central University of Rajasthan, Ajmer, India

E-mail: gaurav.meena@curaj.ac.in

Abstract. The capacity to comprehend and communicate with others via language is one of the most valuable human abilities. We are well-trained in our experience reading awareness of different emotions since they play a vital part in communication. Contrary to popular belief, emotion recognition is a challenging task for computers or robots due to the subjective nature of human mood. This research proposes a framework for acknowledging the passionate sections of conversation, independent of the semantic content, via the recognition of discourse feelings. To categorize the emotional content of audio files, this article employs deep learning techniques such as convolutional neural networks (CNNs) and long short-term memories (LSTMs). In order to make sound information as helpful as possible for future use, models using Mel-frequency cepstral coefficients (MFCCs) were created. It was tested using RAVDESS and TESS datasets and found that the CNN had a 97.1% accuracy rate.

Keywords: Deep learning models, Melfrequency cepstral coefficients, RAVDESS, Convolutional neural networks

1. Introduction

Human-to-human communication through spoken language has been a critical component of civilization from the first settlements as a foundation for data exchange. Similar to the emergence of spoken language as the dominant means of communication, emotions revert to their pre-spoken state [1]. Typically, the term is followed by emotion to make the communication clearer and more understandable for everyone. While confronting emotional predictions, the usual strategy is based on the human brain rather than on technology. Our brains are busy associating all the unique data we will need to communicate with others while focusing on things like how someone appears, the volume and tone of their speech, and so on [2]. Human-machine interaction aims to make computers' human speech understandable. Many instruments nowadays are capable of this, even though it is a signature that sentiments are understood during dialogue to be constantly neglected. Including this perspective in a more comprehensive model of human-machine communication is critical [3]. Hence in this study, we focused on developing a reliable method for identifying the primary emotions sent by a subject during a conversation. Even if each of us may express more than one critical sensation at a time, it isn't easy to discern which and to what degree these sentiments are mixed. As a result of this decision, a model was created to determine just the most powerful emotions in the soundtrack. Machine order feelings have been tried using a variety of approaches, such as PC vision and text evaluation, for example. According to Mel-recurrence cepstral coefficients, we may use unadulterated sound in this work.

The only way to express oneself is via written correspondence. All aspects of a person's body and voice work together to communicate a message successfully. Often, sentiments are expressed by a mix of hand gestures, nonverbal clues, voice, manner, and other physical characteristics. The nonverbal transmission of emotions seems to be universally understood by all peoples of the globe, regardless of how widely spoken their respective languages are. When used in a social environment, these types of new technologies must be able to understand emotional situations in language. Advances in emotion recognition have had a favorable influence on various applications. It is anticipated that automating the emotion identification process will have significant benefits for the disciplines of brain research, psychiatry, and neuroscience, among others. In these divisions of intellectual sciences, a variety of inductions may be generated based on the responses and reactions to a variety of inquiries and circumstances that are applied to the object of research.

A disadvantage may arise if a small number of people are joined together to reflect and ponder whether or not to share. In this approach, the test may benefit from using a PC-based identification framework instead of the traditional methodologies. Discourse-based emotion identification has a wide range of practical applications. In today's world, smart home gadgets and companions are everywhere. To make matters worse, customer service call centers often use automated voice controls that annoy many irate customers. A human chaperon will be able to assist the caller. Individual right hand (Apple Siri) and so forth are only a few examples of many uses. Self-driving cars might be an example of a continuing application. Voice-activated controls are heavily used in these automobiles. Stress, for example, might cause the traveler to speak in foggy terms. Priority must be given to comprehending what the speaker is saying in these situations.

It is necessary to have training datasets to train machine learning algorithms to categorize emotions. There are three kinds of training datasets available for Speech Emotion Recognition tasks: natural, seminatural datasets, and simulated datasets. The natural datasets are derived from publicly accessible movies and audios aired on television or made available online. Aside from contact centers and comparative contexts, there are also databases from other types of situations. Professional voice actors may create seminatural datasets. The simulated datasets, the third and most extensively used form, are comparable to seminatural datasets.

In the following sections, we discover the remainder of the paper: We describe the background research in Section 2; in Section 3, we present a dataset description. Section 4 has an explanation of feature extraction, while Section 5 contains a discussion of several techniques of feature extraction. The findings of the experiments are presented in Section 6. Section 7 presents the conclusion.

2. Literature Review

In recent years, many new articles and studies have been done on speech emotion recognition (SER) [4]. In this part, we examine the most recent investigations that have been conducted in relation to the present work.

From 2000 to 2017, Swain et al. undertook a thorough evaluation of research on SER systems from different perspectives.Traditional machine learning approaches have been used as a classification tool in the study's multiple databases and feature extractions. Deep learning techniques like neural networks were left out, which the authors felt was a mistake. After that, Khalil et al. [5] published a deep learning-based review on SER. Deep neural networks (DNNs) and convolutional neural networks (CNNs) are among the deep learning systems discussed in this study (CNNs). Recurrent neural networks (RNNs), as well as autoencoders, are also discussed in this book. There are several ways to overcome problems that can be implemented immediately. However, this research does not address them. A few weeks ago, Anjali et al. [6] published a review that summarised speech emotion recognition techniques. Speech emotion detection features and approaches from 2009 to 2018 were briefly described and reviewed as part of the inquiry. The extensiveness of the analysis is the paper's primary shortcoming. Nonetheless, it might serve as a starting point for further investigation. Speech emotion datasets and characteristics and noise reduction will be briefly reviewed by Basu et al. [7] in 2020. Finally, they investigate the importance of several classification methods, such as SVM and HMM, in the categorization of speech emotions. An advantage of the research was discovering numerous factors related to SER; nevertheless, its weakness is the leaking of the evaluation of more modern approaches, which only briefly discusses CNN and RNN as a deep learning method. Akcay et al. [8] provide a comprehensive evaluation of databases, characteristics, models, and classifiers of emotions. Classification augmentation methods based on machine learning will also be investigated.

In addition to making significant additions to the previously published works, this research uses deep learning and machine learning approaches to recognize sentiments from speech emotions.

3. Dataset Description

TESS and RAVDESS are two English-language databases that collect recordings of people's feelings when speaking or singing. The dataset's representations are as follows:

3.1. Toronto Emotional Speech Set(TESS)

Experts from the University of Toronto produced an English-language Speech Emotion dataset in 2010 [9]. In all, there are 2800 audio files in the collection that represent discourse expressions in seven different states of excitement: happy, sad, angry, surprised, afraid, disgusted, and neutral. Two performers used the transporter statement "Give the signal.." to convey 200 objective words. It's a performed recording, with actors from two age groups playing out the transcription: 64-year-olds and 26-year-olds.

3.2. The Ryerson Audio-Visual Database of Emotional Speech (RAVDESS)

In this collection, 1440 documents express two lexically-coordinated articulations by 24 skilled entertainers (12 female, 12 male). Every indication is produced at two degrees of emotional power (ordinary, solid), with an added nonpartisan emotion [10][11]. Speech includes calm, happy, sad, furious, afraid, surprise, and disgust articulations. A seven-section mathematical identifier appears in each filename (e.g., 02-01-06-01-02-01-12.mp4). These identifiers designate attributes that provide a boost.

- Methodology (01 = full-AV, 02 = video-just, 03 = sound as it were).
- Vocal channel (01 = discourse, 02 = melody).
- Feeling (01 = nonpartisan, 02 = quiet, 03 = glad, 04 = miserable, 05 = furious, 06 = unfortunate, 07 = disdain, 08 = astounded).
- Passionate power (01 = ordinary, 02 = solid). NOTE: There is no solid force for the 'nonpartisan' feeling.
- Explanation (01 = "Children are talking by the door", 02 = "Canines are perched by the entryway").
- Redundancy (01 = first reiteration, 02 = second redundancy).
- Entertainer (01 to 24. Odd numbered entertainers are male, even numbered entertainers are female)

4. Feature extraction

A speaker's emotional state is derived from speech patterns in speech emotion recognition [12]. The voice signal's acoustic characteristics are highlighted. Feature extraction is the process of extracting a small amount of data from a voice signal so that it may be used to address each speaker individually. Prosodic and Phonetic elements of a speech percept are the two primary groups from which the emotional content of a speech percept may be constructed. Among the prosodic characteristics are pitch and loudness and energy, and formant. The phonetic elements of a language are typically associated with the way words are spoken. As a result, the inquiry focuses on prosodic traits or a combination of them to help people recognize emotions. When conveying an emotional message, pitch and volume are often crucial.

4.1. List if features

In this study, we employed Mel Frequency Cepstral Coefficients (MFCC) and Mel Spectrogram, which were derived from the audio data. For feature extraction, we turned to the Librosa package's Python implementation.

- When considering a signal, the MFCC is a short group of characteristics that succinctly depict a spectral envelope's basic shape.
- Mel spectrogram plots sufficiency on recurrence versus time chart on a "Mel" scale.

4.2. Mel Frequency Cepstral Coefficient

For example, echoes show prominent peaks in the recurrence spectrum in the usual analysis of time signals. Any cepstrum feature is obtained by using Fourier Transform on a spectrogram. Because it is based on the Mel scale, a tone's apparent repetition or frequency may be directly related to its actual estimated frequency, an outstanding feature of MFCC. It adjusts the frequency to better match what the human ear can detect. The term "Mel" refers to the frequency-to-pitch metering scale. The following formula may be used to convert an estimated value in frequency scale to a Mel scale value [13].

$$m = 2595 \log 10(1 + \tfrac{f}{700}) \quad (1)$$

4.3. Melspectrogram

A spectrogram is a time-varying graphic depiction of the frequency spectrum of sound or other phenomena. It's a visual representation of how frequencies change over time.

4.4. Data Visualization

The problem and the kind of solution may both be better understood by visualizing the dataset. When working with unprocessed audio, graphic representations become more important to make sense of the information. Listening to audio samples or sound clips may be time-consuming and need a lot of focus, but visualizing what you're hearing can help establish context rather quickly. An example of data visualization is shown in figure 1.

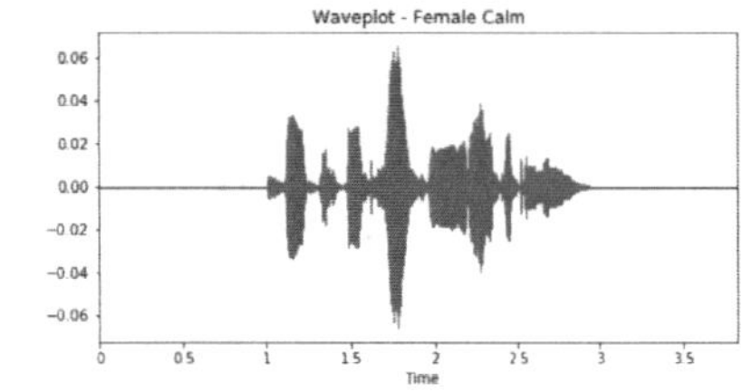

(a) Women: Calm

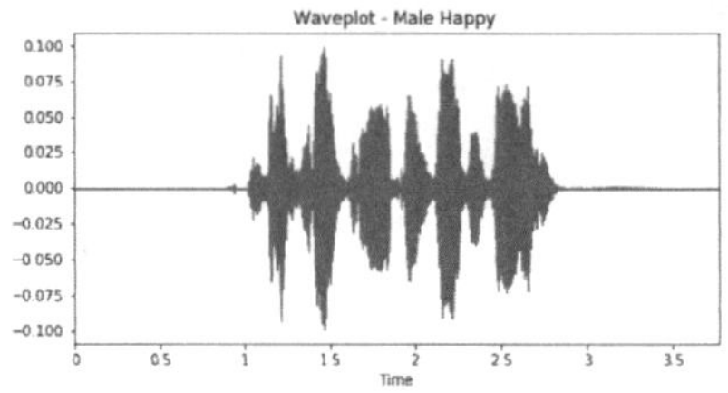

(b) Man: Happy

Figure 1: Data visualization

A man and a female were compared to say identical words in anger. Males and females are unique in their voices since women tend to have a higher pitch. Figure 2 depicts the scenario.

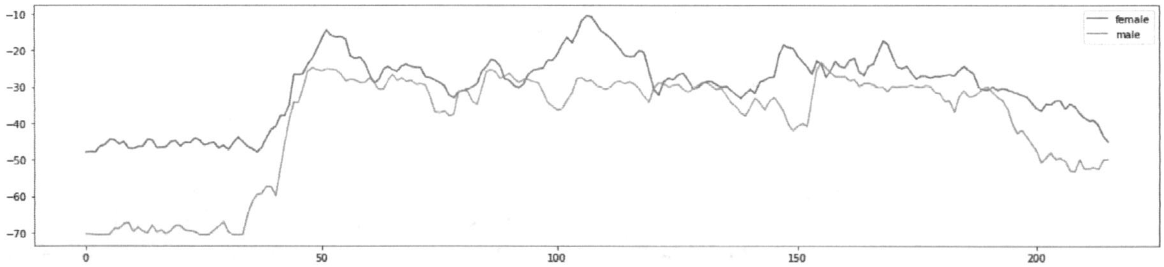

Figure 2: Angry: Male, Female

4.5. Data Augmentation

When we don't have enough data or need to add it to our dataset, a common approach is to use data augmentation to provide diversity. Minor changes to the current data samples are all it takes to get the job done. Data augmentation is an essential tactic for deep learning systems [5]. Enhancement in speech recognition entails altering the sound waveform used in training in some way, such as by accelerating or slowing spoken audio and adding background noises. When numerous enhanced copies of the same input are delivered into the network during training, the dataset grows. The following are data augmentation methods.

4.5.1. Noise Addition Noise, such as white noise, is added to the sample in this approach. Random samples having mean value 0 and a standard deviation value 1 are known as white noises.

4.5.2. Time Stretching In this technique, the sound is sped up or slowed down without altering the pitch. librosa's *time_stretch* function may be used to do this. Time stretching accepts wave samples and a factor for stretching as inputs to the *Time_stretch* function.

4.5.3. Pitch Shifting The pitch of the sound is altered without influencing the speed of the sound. librosa's $pitch\_shift$ function is used for it. Wave samples, sampling rate, and the number of pitch steps are all considered.

Figure 3 depicts a wave plot of a sound file with noise added, time stretching, and pitch shifting.

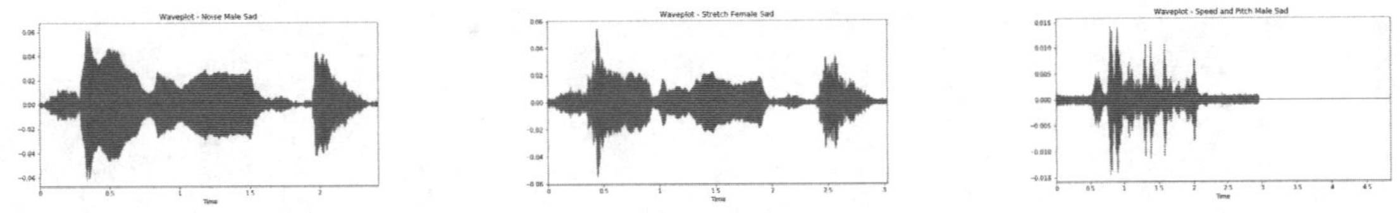

(a) Added noise (b) Time stretching (c) Pitch shifting

Figure 3: Wave plot with Data Augmentation

5. Methods Used

5.1. Deep Learning

The suggested model for classifying emotions is based on a deep learning method using CNNs, LSTMs, and GRUs.

5.1.1. CNN CNN's are one of the most well-known deep learning models that have demonstrated great success in object detection, handwriting recognition, voice recognition, and NLP [1]. The numerical activity of convolution is used in these organizations, thus the word convolution. Convolutional, pooling, and associated layers make up the majority of CNNs' building components (Fully Connected). The accuracy, loss, and confusion matrix of the CNN model are shown in figure 4 below.

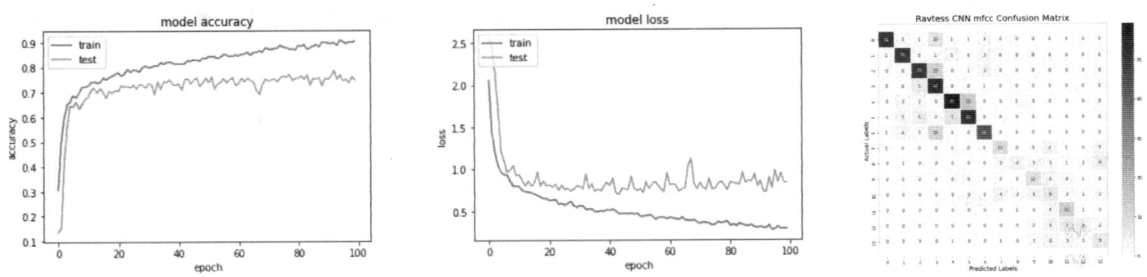

(a) Accuracy (b) Loss (c) Confusion matrix

Figure 4: CNN model

5.1.2. LSTM Because of their input connection, RNN may learn and adapt to the temporary occasion without modifying the progressively molded loads. Uses in which time is critical, such as speech processing, music arrangement, and video representation, may benefit from this. In any event, error signals flowing backward in time created via Backpropagation over time may either increase in size or disappear depending on the amount of the loads. This will either cause loads to swing or postpone the preparation and merger of the organization. LSTM avoided the problems raised above by combining RNN short-term adaptation. LSTM networks can handle data successions that are both incompressible and noisy, even over periods exceeding 1000 stages. As a result, they may shorten the inclination computations at a fixed location without affecting the lengthy enactments. LSTM networks have recently focused on various applications, such as time-series events. Two of these applications are Speech Processing and Emotion Detection. The LSTM model's accuracy, loss, and confusion matrix are shown in figure 5 below.

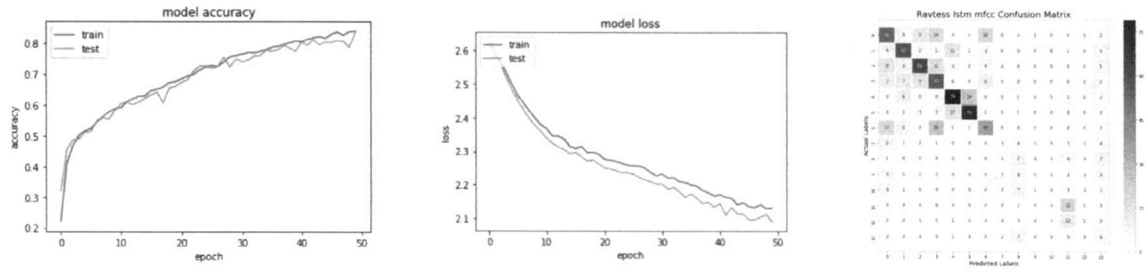

(a) Accuracy (b) Loss (c) Confusion matrix

Figure 5: LSTM model

5.1.3. GRU GRU is a more advanced version of the Recurrent Neural Network. Circle-shaped RNNs are those that keep on going with the data. Long-reach circumstances and temporal dependencies make the GRU network appropriate for speech emotion identification. The network has been used to distribute the initial tests. The mistake is propagated back through the system, in the same manner it was introduced to alter the variables in the original and obtained outputs. The components in this interaction are repeatedly re-examined until they are clear. These factors are used to construct a prediction as new information becomes available. Because there are fewer parameters in GRU networks, they run faster. It can manage the data stream from the previous enactment and expose the complete memory to the network. Reset and update gates r and z are used in the GRU network. Figure 6 illustrates the GRU model's accuracy, loss, and confusion matrix.

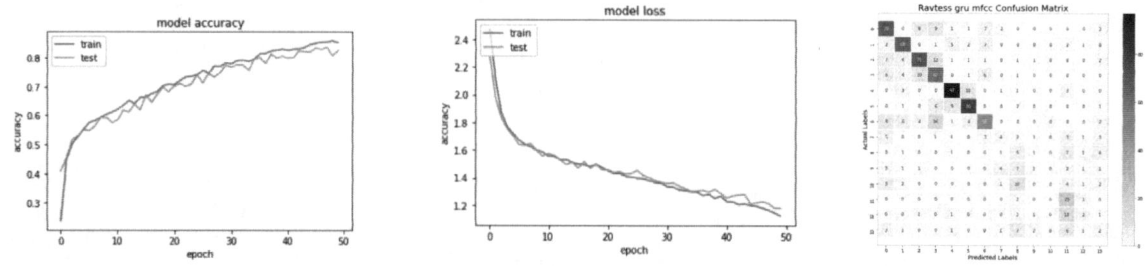

(a) Accuracy (b) Loss (c) Confusion matrix

Figure 6: GRU model

5.2. Machine Learning
5.2.1. SVM Sub classification and regression may be performed using SVMs. Support vectors, physical points in space, are constructed using a data point's attributes. The feature vector values associated with each data point in an N-dimensional dataset are shown as coordinate points on an N-dimensional region. Classification is accomplished by locating a hyperplane in outer space that clearly delineates the different categories. SVM excels in high-definition voice emotion recognition using 40-dimension data. SVM implementation is impossible without first locating the hyperplane [14][15]. Two conditions must be met to choose the proper hyperplane.

- The hyperplane should be the most accurate in classifying the classes.
- It is necessary to increase the margin distance between the hyperplane and the nearest data point.

Journal of Physics: Conference Series **2236** (2022) 012003 doi:10.1088/1742-6596/2236/1/012003

Low-dimensional data may be transformed into high-dimensional data using the kernel technique in SVM, which provides additional properties that allow the identification of hyperplanes. Linear SVM is a more efficient SVM implementation in terms of computation time. Table 1 illustrates the computing performance of the SVM model on Tess dataset.

Table 1: Computation performance of SVM

Precision(%)	Recall (%)	Accuracy (%)	F1-score (%)
86	86	86.15	86

5.2.2. Random Forest Similar to decision trees, random forests are supervised categorization methods. While the decision tree uses Gini and information gain values to determine the root node and splitting features, the random forest method generates these values at random. There are many trees in the random forest; thus, the more trees we have, the better we get. Random forests may suffer from overfitting. However, increasing the number of trees can help. Other features of the random forest include its capacity to manage missing data and categorize categorical variables with many classes [10]. The following are the stages involved in creating a random forest classifier:

(i) Choose a subset of features from the dataset to work with.
(ii) Choose a node from the subset of characteristics that have been picked using the best split approach.
(iii) Continue using the best split approach to create child nodes from the subset of characteristics that have been selected.
(iv) Continue in this manner until all nodes have been utilized as splits.
(v) Create a forest by iteratively creating several trees using steps 1-4 to make a forest.

The higher the number of trees in the Random Forest, the more time it takes to develop the forest. Table 2 illustrates the computing performance of the random forest model on Tess dataset.

Table 2: Computation performance of Random forest

Precision(%)	Recall (%)	Accuracy (%)	F1-score (%)
96	96	96	96

6. Experiments

6.1. Experimental setting

We are now using deep learning and machine learning models to identify voice utterances based on emotional content. We employed an emotional database to train and assess our models in addition to three commonly used standards for emotion identification from voice utterances. TensorFlow and Keras were the frameworks we utilized to implement our models. As a preview of our current work, we've included this part. The following libraries are being used:

- Soundfile: Soundfile is a Python package to read the sound file of various formats, for instance, WAV, FLAC, OGG, MAT files.
- Librosa: Librosa is a Python package for sound and music examination, for instance, feature extraction and manipulation, segmentation, visualization, and display.
- Os: Os is a Python package for utilizing a working framework, for instance, acquire the base name of a file, open the record in various modes like reading, write, append.

- Glob: Glob is a Python package for discovering path or path names of the documents and files, the record having some particular pattern, For instance, all files of.WAV extension.
- Numpy: Numpy is a Python package for logical estimation, for instance performing various operations or procedures on matrices.
- Sklearn: Sklearn is a Python package for performing distinctive machine learning operations, for instance predicting the unknown future values.

6.2. Result discussion

Numerous Algorithms for Speech Emotion Recognition have been developed [5-7]. We compared our method to other best-in-class strategies in Table1. According to the data in Table 3, our suggested approach surpasses the other three standard methods. The model's performance on the RAVDESS and TESS datasets was evaluated and found to be adequate.

Table 3: Performance comparison

Authors	Year	Database	Accuracy (%)
Zeng et al.[15]	2019	Ravdess	64.48
Zamil et al.[16]	2019	Ravdess	67.14
Issa et al.[17]	2020	Ravdess	71.61
Our Model (CNN)	2022	Ravdess	**72.2**
Dupuis et al. [18]	2011	Tess	82
Praseetha et al.[19]	2018	Tess	95.82
Huang et al.[20]	2019	Tess	85
Our Model (CNN)	2022	Tess	**97.1**

7. Conclusions

Using the RAVDESS and TESS datasets, we proposed methods based on deep neural networks and machine learning to categorize emotions. The overall accuracy of 0.97 was achieved by training the models to classify unique emotions. The sad class had the most outstanding performance, with a score of 0.90, and the worst on the surprised and disgust class (0.77). MFCC characteristics (spectrum of a-spectrum) were retrieved from the audio samples utilized for training. When comparing SVM with random forest, we found an accuracy of 0.86 for SVM and 0.96 for the random forest, respectively. We achieved an accuracy of 0.92, 0.93, and 0.971, respectively, using the deep learning model LSTM, GRU, and CNN. We may conclude that deep neural networks constitute a solid foundation for tackling the problem despite the positive outcomes. The suggested work's efficiency, accuracy, and usefulness may all be improved in the future. Mood swings and despair are two more possible additions to the model's capabilities. Therapists may use these methods to keep track of their patients' mood fluctuations. Adding a sarcasm detection system to a computer with emotions is difficult.

References

[1] Sajjad, Muhammad, and Soonil Kwon. "Clustering-based speech emotion recognition by incorporating learned features and deep BiLSTM." IEEE Access 8 (2020): 79861-79875.
[2] Sonmez, YeSim Ülgen, and Asaf Varol. "New trends in speech emotion recognition." 2019 7th International Symposium on Digital Forensics and Security (ISDFS). IEEE, 2019.
[3] Huang, Kuan-Chieh, and Yau-Hwang Kuo. "A novel objective function to optimize neural networks for emotion recognition from speech patterns." 2010 Second World Congress on Nature and Biologically Inspired Computing (NaBIC). IEEE, 2010.
[4] Abbaschian, Babak Joze, Daniel Sierra-Sosa, and Adel Elmaghraby. "Deep learning techniques for speech emotion recognition, from databases to models." Sensors 21.4 (2021): 1249.

[5] Khalil, Ruhul Amin, et al. "Speech emotion recognition using deep learning techniques: A review." IEEE Access 7 (2019): 117327-117345.

[6] Tripathi, Anjali, et al. "A Review on Emotion Detection and Classification using Speech." Proceedings of the International Conference on Innovative Computing Communications (ICICC). 2020.

[7] Basu, Saikat, et al. "A review on emotion recognition using speech." 2017 International conference on inventive communication and computational technologies (ICICCT). IEEE, 2017.

[8] Akçay, Mehmet Berkehan, and Kaya Oğuz. "Speech emotion recognition: Emotional models, databases, features, preprocessing methods, supporting modalities, and classifiers." Speech Communication 116 (2020): 56-76.

[9] Dupuis, Kate, and M. Kathleen Pichora-Fuller. "Toronto emotional speech set (TESS)-Younger $talker\_Happy$." (2010).

[10] Arya, Resham, et al. "Speech based Emotion Recognition using Machine Learning." 2021 IEEE Mysore Sub Section International Conference (MysuruCon). IEEE, 2021.

[11] Livingstone, Steven R., and Frank A. Russo. "The Ryerson Audio-Visual Database of Emotional Speech and Song (RAVDESS): A dynamic, multimodal set of facial and vocal expressions in North American English." PloS one 13.5 (2018): e0196391.

[12] Aldeneh, Zakaria, and Emily Mower Provost. "Using regional saliency for speech emotion recognition." 2017 IEEE international conference on acoustics, speech and signal processing (ICASSP). IEEE, 2017.

[13] Yang, Ningning, et al. "Recognize basic emotional statesin speech by machine learning techniques using mel-frequency cepstral coefficient features." Journal of Intelligent & Fuzzy Systems Preprint (2020): 1-12.

[14] Deshmukh, Girija, et al. "Speech based emotion recognition using machine learning." 2019 3rd International Conference on Computing Methodologies and Communication (ICCMC). IEEE, 2019.

[15] Zeng, Yuni, et al. "Spectrogram based multi-task audio classification." Multimedia Tools and Applications 78.3 (2019): 3705-3722.

[16] Zamil, Adib Ashfaq A., et al. "Emotion detection from speech signals using voting mechanism on classified frames." 2019 International Conference on Robotics, Electrical and Signal Processing Techniques (ICREST). IEEE, 2019.

[17] Issa, Dias, M. Fatih Demirci, and Adnan Yazici. "Speech emotion recognition with deep convolutional neural networks." Biomedical Signal Processing and Control 59 (2020): 101894.

[18] Dupuis, Kate, and M. Kathleen Pichora-Fuller. "Recognition of emotional speech for younger and older talkers: Behavioural findings from the Toronto emotional speech set." Canadian Acoustics 39.3 (2011): 182-183.

[19] Praseetha, V. M., and Sangil Vadivel. "Deep learning models for speech emotion recognition." Journal of Computer Science 14.11 (2018): 1577-1587.

[20] Huang, Andrew, and Puwei Bao. "Human vocal sentiment analysis." arXiv preprint arXiv:1905.08632 (2019).

Deep learning for facial emotion recognition using custom CNN architecture

Mr. Rohan Appasaheb Borgalli[1], **Dr. Sunil Surve**[2]
[1] Research Scholar, Department of Electronics Engineering
[2] Professor, Department of Computer Engineering
Fr. Conceicao Rodrigues College of Engineering, Bandra, Mumbai University, Mumbai, India

Email: rohanborgalli111@gmail.com

Abstract. Human facial expressions are an indication of true emotions. To recognize facial expressions accurately is useful in the field of Artificial Intelligence, Computing, Medical, e-Education, and many more. The facial expression recognition (FER) system detects emotion through facial expression. But, it is challenging to detect facial emotions accurately. However, recent advancements in technology, research, and availability of facial expression datasets have led to the development of many FER systems which can accurately detect facial emotions. Past research in the field of FER indicates With Convolutional Neural Networks (CNNs), deep learning techniques are the most advanced presently. Custom CNN Architecture is used to implement basic facial emotion recognition in static images in this paper. A K-fold cross-validation method was used to train them using FER13, CK+, and the JAFFE data set. On the seven classes of fundamental emotions, including anger, disgust, fear, happiness, neutrality, sorrow, and surprise, the FER13, CK+, and JAFFE datasets had an accuracy rate of 91.58 percent. Given the difficulty of developing unique CNN architecture, this study's accurate findings contrast well with those of previous studies.

1. Introduction

Interpersonal communication depends on facial expressions to convey a person's emotional state, cognitive activity, purpose, personality, and psychopathology. Over the last several decades, it has been the subject of much research. Facial expression recognition is still a challenge despite the progress that has been achieved because of the complexity and variety in facial expressions.

In general, nonverbal signals like hand gestures, facial expressions, and unconscious speech may be used to communicate our thoughts and feelings. This nonverbal communication system has the potential to be very beneficial. What matters most is how well the facial expression recognition (FER) system recognizes or extracts an image's facial expression. Many sectors of application might benefit from the technology, including lie detection, medical evaluation, and human-computer interaction.

In the past, researchers have addressed FER using methods like the Multi-layer Perceptron Model, k-Nearest Neighbors, and Support Vector Machines. To extract information, these algorithms utilized Local Binary Patterns, Eigenfaces, Face-Landmark features, and Texture features. Neural networks have acquired the most traction among these techniques, and they are widely used for FER. CNN's have lately acquired favor in deep learning due to their laid-back character and ability to produce decent results without the need for human feature extraction from raw image data. Learning high-level features from training data is a goal of deep learning algorithms. This is a unique feature of Deep

Learning, and it puts it one step ahead of more typical Machine Learning techniques. Deep learning minimizes the load when it comes to discovering new characteristics for each challenge. Deep learning has a strong benefit over regular machine learning when it comes to performance as the amount of data grows. When there is a lack of data, deep learning algorithms fail. In order to properly train deep learning algorithms, a large amount of data is required.

There are three main face emotion datasets: FER13, JAFFE, and K-fold cross-validation. In this research, we use deep learning CNNs to build an architecture that may be used to build an effective facial emotion recognition system.

2. Literature Survey

Since the advent of Deep Learning in the field of Machine Vision [1], which uses a CNN [2] for feature extraction, classification, and recognition tasks, state-of-the-art results have been achieved in a variety of fields, including scene comprehension, object identification, and face recognition.[3],[4].

Deep CNN-based architecture extracts feature themselves and are useful for image classification [5]. Such features provide application-specific learning over traditional hand-crafted image features [6] which are having limitations. Transfer learning is a popular term for this method. To put it another way, the capacity of a system to adapt what it has learnt in one activity to another is called "transfer learning" [7]. Ramalingam [8] suggested transferring VGG19-based learning to the FER2013 dataset. For new datasets such as CK+ or JAFFE, he used what he learned from FER2013 to the CK+ and JAFFE databases. Caramihal et al. suggested a new FER system based on a Generative Adversarial Network (GAN) classifier.[9]. In contrast to the traditional FER, the new GAN has 14 subclasses for each emotion, with a real class and a fake class for each. The GAN classifier is generally trained using N-classes, while this technique uses 2N-classes to train the GAN classifier. Increased detection accuracy was a result.

Facial Action Coding Units (AUs) learned with the CK+ Dataset are used to train a Deep Learning model-based CNN that uses extracted features to classify emotions. FER-2013, CK+, JAFFE, and FERG datasets, Minaee et al. [11] suggested deep learning based on visualization approach attentional convolutional network, which can concentrate on significant areas of the face to find face regions for recognizing various emotions.

This paper focused on CNN-based efficient architecture to correctly recognize basic human emotion (happy, sad, fear, disgust, anger, surprise, and neutral) on existing datasets to build robust and efficient Facial Expression Recognition (FER) systems.

3. Dataset Overview

The fundamental emotion dataset has seven key categories: anger, contempt, fear, pleasure, sadness, surprise, and neutral. For CK+, there is also a separate contempt class.

In a Kaggle recognition competition, the Fer2013 dataset was obtained (Facial Expression Recognition Challenge). Table 1 presents the distribution of each dataset obtained from [12] for each class. Facial expression recognition works well with Fer2013.

Table 1. Distributions of Every Class in FER13, CK+ and JAFFE Dataset [12]

	FER13	CK+ (Last Frame)	JAFFE
Angry	4,593	45	30
Disgust	547	59	30
Fear	5,121	25	31
Happy	8,989	69	31
Sad	6,077	28	31
Surprise	4,002	83	30
Neutral	6,198	327	30
Contempt	0	18	0

3.1. FER dataset [13]

The Fer2013 dataset was retrieved in a Kaggle recognition competition (Facial Expression Recognition Challenge). Grayscale images of female faces of various ages make up the dataset. Each image has a resolution of 48x48 pixels. This data is divided into seven subcategories (angry, disgust, fear, happy, sad, surprise, and neutral). Fer2013 has a total of 28,709 images, 3,589 of which are for validation and 3,589 for testing. Faces of all ages and in all directions are included in the dataset. Facial expression recognition works very well Fer2013. Figure 1 Shows FER2013 Dataset Sample Images.

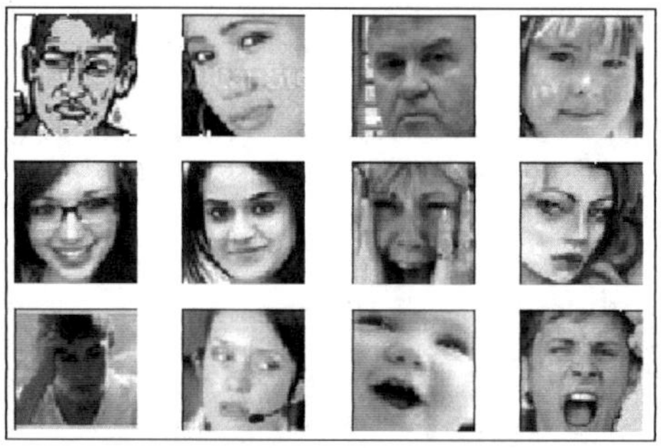

Figure 1. FER2013 Dataset Sample Images

3.2. Japanese Female Facial Expression (JAFFE)Dataset [14]

Seven Japanese female models posed for 213 grayscale images of their faces in seven different facial emotions (angry, disgusted, neutral, sad, surprised, joyful, and fearful). Each image has a pixel size of 256x256. In order to construct the JAFFE dataset, scientists used data from three different sources. The training set has 129 images, the test set contains 42 images, and the validation set contains 42 images (20 percent). Kyushu University's Michael Lyons, Miyuki Kamachi, and Jiro Gyoba collaborate to develop it. Most facial expression recognition tasks and traditional facial expression databases are based on this dataset. The suggested algorithm's input layer has a 48x48x1 image size thus we first standardized the dimensions of all images to 48x48x1. Figure 2 Shows JAFFE Dataset Sample Images.

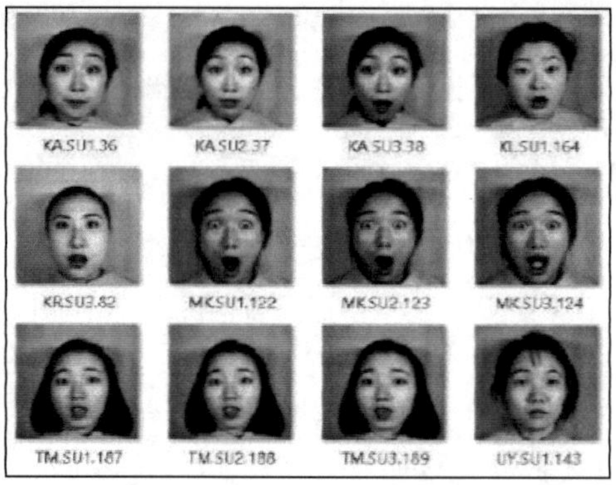

Figure 2. JAFFE Dataset Sample Images

Journal of Physics: Conference Series **2236** (2022) 012004 doi:10.1088/1742-6596/2236/1/012004

3.3. CK+ dataset [15]

Database for Cohn-Kanade Plus (CK+) An additional 22% and 27% of sequences and persons were added to the Cohn-Kanade (CK) dataset in 2010 to form the CK+ dataset. The peak frames of 327 image sequences were applied with FACS-coded emotion labels from neutral to peak emotions of both posing and non-posing (spontaneous) expressions. The database's 123 participants range in age from 18 to 50, with 79% of them being female and 81% being European-American, 13% being Afro-American, and 6% being from other ethnicities. Contempt is added to this dataset's eight facial expression classifications, making this dataset the most comprehensive yet. But as we are focusing on basic 7 emotions then as contempt emotion is near to disgust, we converted contempt labels to disgust and considered only 7 basic emotions of CK+ Dataset. Figure 3 Shows CK+ Dataset Sample Images.

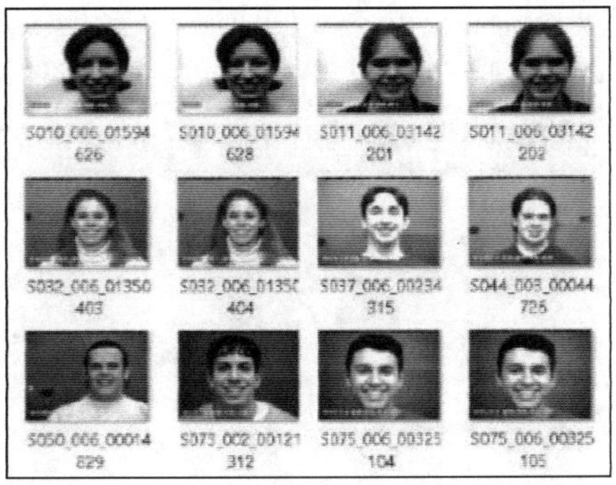

Figure 3. CK+ Dataset Sample Images

4. Proposed Methodology

The proposed methodology is based on a Convolution Custom CNN Architectures with an input layer and some Convolutional layers are used to train in this neural network-based deep learning framework. Edges, blobs, hues, and last layers are examples of low-level features. In our case, it's the Basic Emotions like Angry, Disgust, Fear, Happy, Neutral, Sad, and Surprise that certain fully-connected layers and an output layer learn.

Basic Emotion databases such as FER13, CK+, and JAFEE have a variety of images but to learn features faster we are preprocessing Images so only the required part of the face is captured using the Haar Cascade algorithm for detection of frontal face.

The training datasets consist of different basic facial Emotions. Training data, on the other hand, tend to be unevenly distributed. Network performance will suffer if there is a discrepancy in the distribution of training data. Because of this negative impact, we supplemented the data distribution by performing data generation (shown in Figure 6) and a Data Augmentation operation (as shown in Figure 6). It is possible to reduce the cross-entropy loss function using the Adam optimizer.

Before training a machine learning model, resampling approaches are used to verify the model's accuracy using a small data set. K-fold Cross-Validation is one of these methods. We used K=10 for our case and accordingly trained the model and tested it for 7 Basic Emotions. The proposed methodology for basic emotion detection using custom CNN architecture is shown in Figure 4.

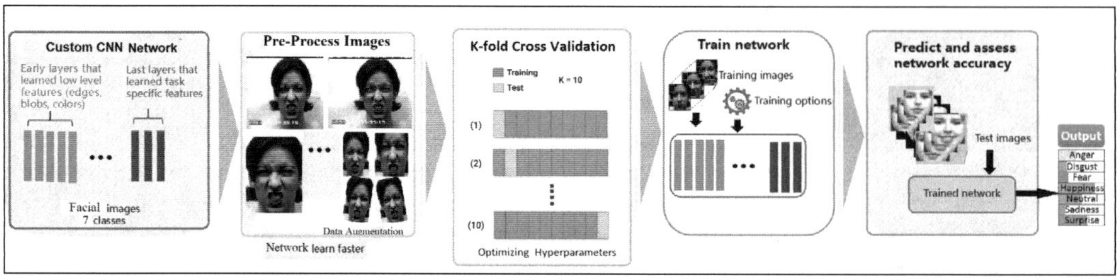

Figure 4. Proposed methodology for Basic Emotion detection using Custom CNN Architecture

The proposed Custom CNN uses many convolutional layers, as well as a few layers that mix maximum and average pooling, and a small number of fully-connected layers, to process a 48x48x3 image. This model, which has six convolutional layers and 1.48M parameters, uses Rectified Linear Units (ReLU) and batch normalization. Fig. 5 and Fig. 6 show the Customized CNN Architecture architectural detail.

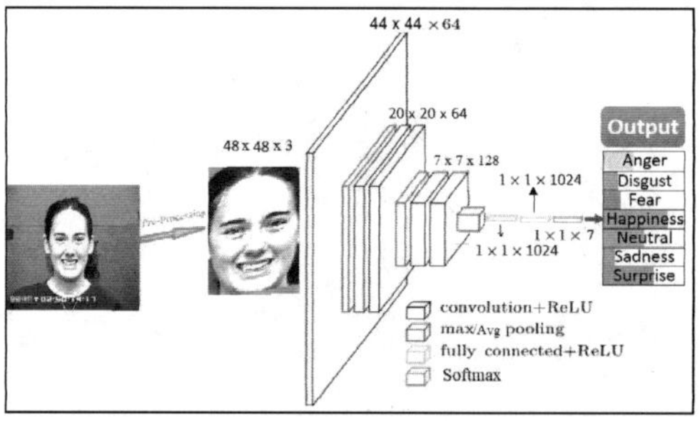

Figure 5. Architectural detail of Custom CNN Architecture used for FER13, CK+ & JAFFE

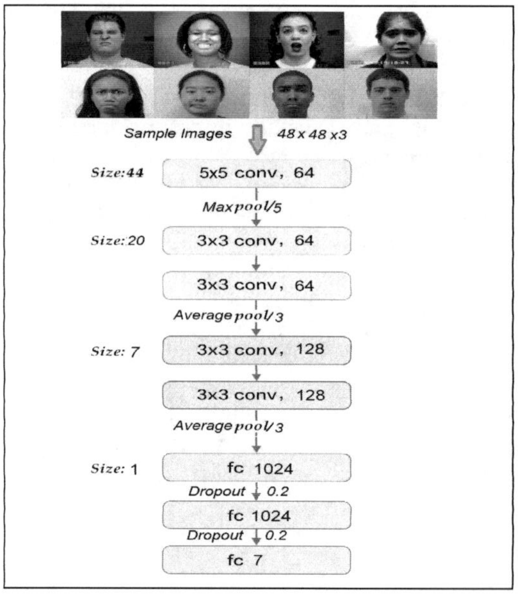

Figure 6. Architectural detail of Custom CNN Architecture used for FER13, CK+ & JAFFE

4.1. K-fold cross-validation technique [16]

For the K-fold cross-validation technique, the k parameter determines how many groups a given data sample must be divided into. K-fold cross-validation refers to this method. If a specified number for k is selected, then it is mentioned in place of K. For the experiment, we used k=10, which is 10-fold cross-validation.

Cross-validation is a machine learning technique used to assess a model's capacity to adapt to new data. It is possible to test the performance of the model with data that was not included in the training phase using a small sample.

One of the most often used strategies is a simple train-to-test split since it's easy to grasp and provides a more accurate estimate of the model's skill than other ways, such as the basic split.

The general procedure is as follows:

1. Randomly shuffle the entire dataset and divide it into training and testing groups.
2. Take a look at the training data and split it into k groups.
3. For each unique group of training data:
   (a). As a holdout or data set for testing, use the group.
   (b). Use the remaining groups as a training data set.
   (c). Make a model out of the training data by comparing it to the test data.
   (d). After each cycle, the evaluation score is saved, and the model is removed.
4. Using a sample of model assessment scores, describe the model's performance.
5. Each observation in the data sample belongs to a single group and remains with that group all through the procedure. This means that each sample can be utilized once in the hold outset and once in the model training (k-fold)
6. Finally, we test it on a test set using the best-trained model on the training set, which is further evaluated to get an accuracy score.

The complete procedure of K-fold Cross-validation is shown in Figure 7.

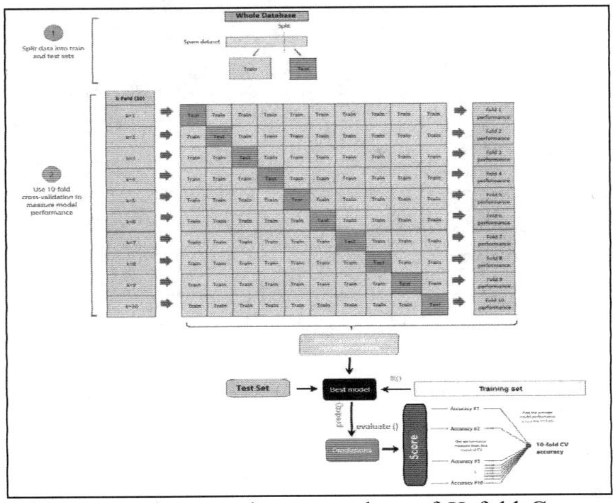

Figure 7. The complete procedure of K-fold Cross-validation

5. Results

5.1. MultiClass Confusion Matrix

For each item, there are three columns: true/actual classification and predicted classification, and these three values are recorded in the confusion matrix. In order to keep things consistent, columns are used to indicate model predictions, while rows are used to represent the actual classifications. As a result,

Journal of Physics: Conference Series **2236** (2022) 012004 doi:10.1088/1742-6596/2236/1/012004

correctly categorized items may be found on a diagonal from top left to bottom right, which corresponds to the number of times each rater agreed with each other. The accuracy of the actual labels compared to those predicted is being assessed using a confusion matrix.

5.2. Accuracy Score

An f1-score is produced for each class of test samples to measure the model's accuracy. Indicating the overall accuracy as well as the accuracy for each class predicted.

5.2.1. Accuracy Score and Confusion Matrix of FER13 Database

Seven basic face expressions are included in the FER13 Database: Angry, Disgust, Fear, Happy, Sadness, Surprise, and Neutral (see also FER12 Database). It seems that the most of the classes have been categorised correctly, except for fear, in the Confusion Matrix. For some people, fear is mistaken for sadness. Indicating an overall accuracy of 85 percent, the accuracy score reflects the precision, recall, and f1-score for each test sample. Figure 8 depicts the FER13 Database's adjusted Confusion Matrix.

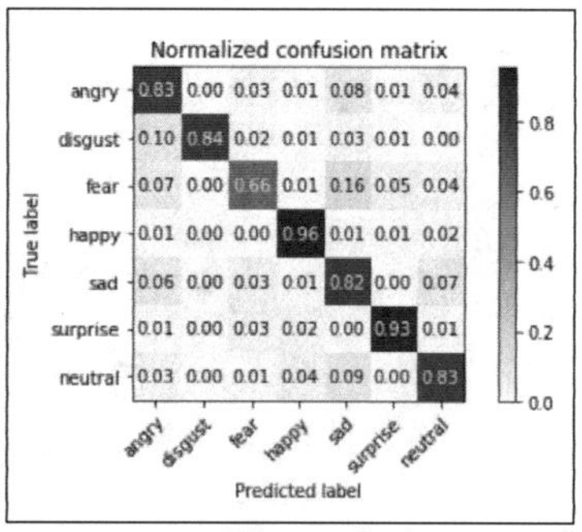

Accuracy Score :				
	precision	recall	f1-score	support
0	0.79	0.83	0.81	411
1	0.92	0.84	0.88	417
2	0.85	0.66	0.74	451
3	0.94	0.96	0.95	805
4	0.73	0.82	0.77	600
5	0.89	0.93	0.91	370
6	0.84	0.83	0.83	545
accuracy			0.85	3589
macro avg	0.85	0.84	0.84	3589
weighted avg	0.85	0.85	0.84	3589

Figure 8. Accuracy Score and Confusion Matrix for FER13 Database

5.2.2. Accuracy Score and Confusion Matrix of CK+ Database

Angry/Disgust/Fear/Sad/Surprise/Neutral are all included in the CK+ Database's 7 Basic Facial Emotions. Except for fear, it appears that the most of the classes in the Confusion Matrix have been accurately categorized. For some people, fear is mistaken for sadness. For each class, the Accuracy score represents precision, recall, and f1-score. With an overall accuracy rate of 81 percent. Figure 9 shows the normalized Confusion Matrix for CK+ Database.

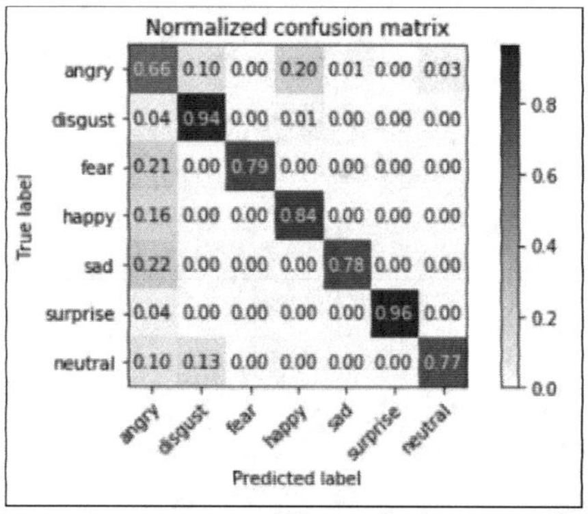

Accuracy Score :				
	precision	recall	f1-score	support
0	0.64	0.66	0.65	147
1	0.77	0.94	0.85	69
2	1.00	0.79	0.89	97
3	0.69	0.84	0.75	79
4	0.95	0.78	0.85	49
5	1.00	0.96	0.98	107
6	0.86	0.77	0.81	39
accuracy			0.81	587
macro avg	0.84	0.82	0.83	587
weighted avg	0.83	0.81	0.81	587

Figure 9. Accuracy Score and Confusion Matrix for CK+ Database

5.2.3. Accuracy Score and Confusion Matrix of JAFFE Database

Anger, disgust, fear, pleasure, sorrow, surprise, and neutrality are all reflected in the JAFFE Database. It seems that the most of the classes have been classified correctly, except for fear, in the Confusion Matrix. For some people, fear is mistaken for sadness. For each class, the Accuracy score indicates precision, recall, and f1-score. Indicating a 95% accuracy rate. Figure 10 shows the normalized confusion matrix for the JAFFE database.

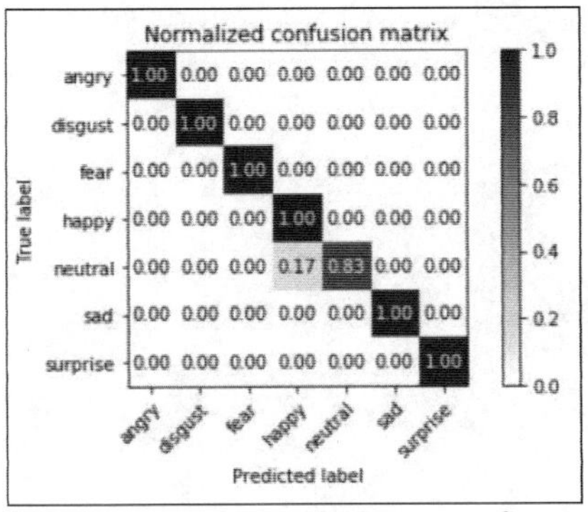

Accuracy Score :	precision	recall	f1-score	support
0	1.00	1.00	1.00	1
1	1.00	1.00	1.00	3
2	1.00	1.00	1.00	3
3	0.83	1.00	0.91	5
4	1.00	0.83	0.91	6
5	1.00	1.00	1.00	2
6	1.00	1.00	1.00	1
accuracy			0.95	21
macro avg	0.98	0.98	0.97	21
weighted avg	0.96	0.95	0.95	21

Figure 10. Accuracy Score and Confusion Matrix for JAFFE Database

5.3. Emotion Analysis on Sample Test Image

Emotion Analysis of Sample test Images is done at the end to check the working of the model from FER13, CK+, and JAFFE Dataset. Figure 11-13 shows for sample image with softmax output indicating a probability of each emotion with plotted graph the maximum will be considered as Predicted emotion by the proposed model from FER13, CK+, and JAFFE Dataset.

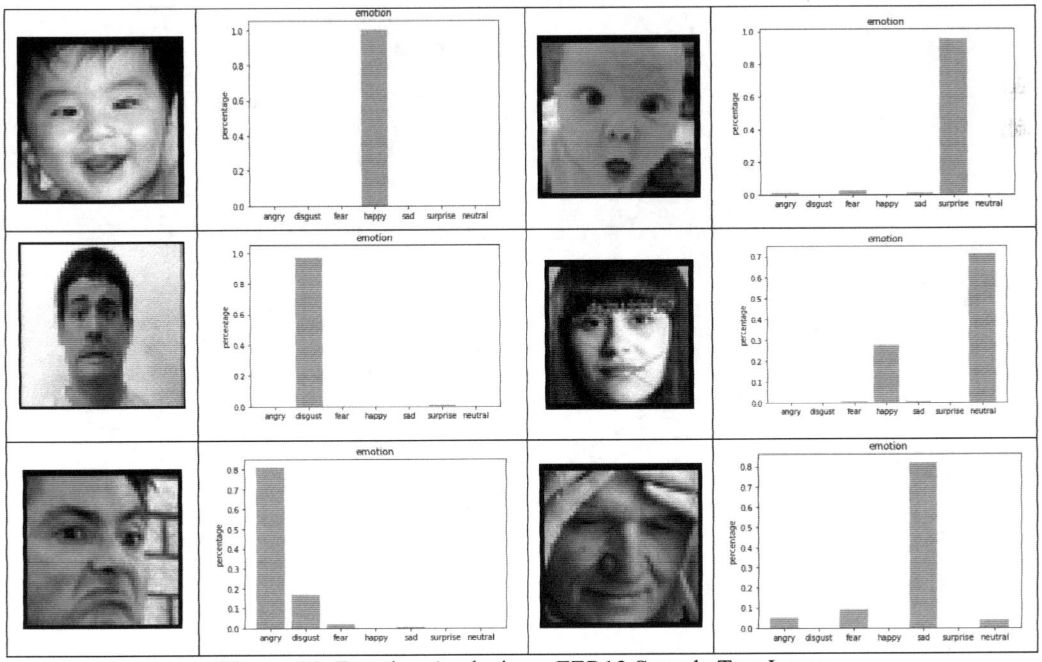

Figure 11. Emotion Analysis on FER13 Sample Test Image

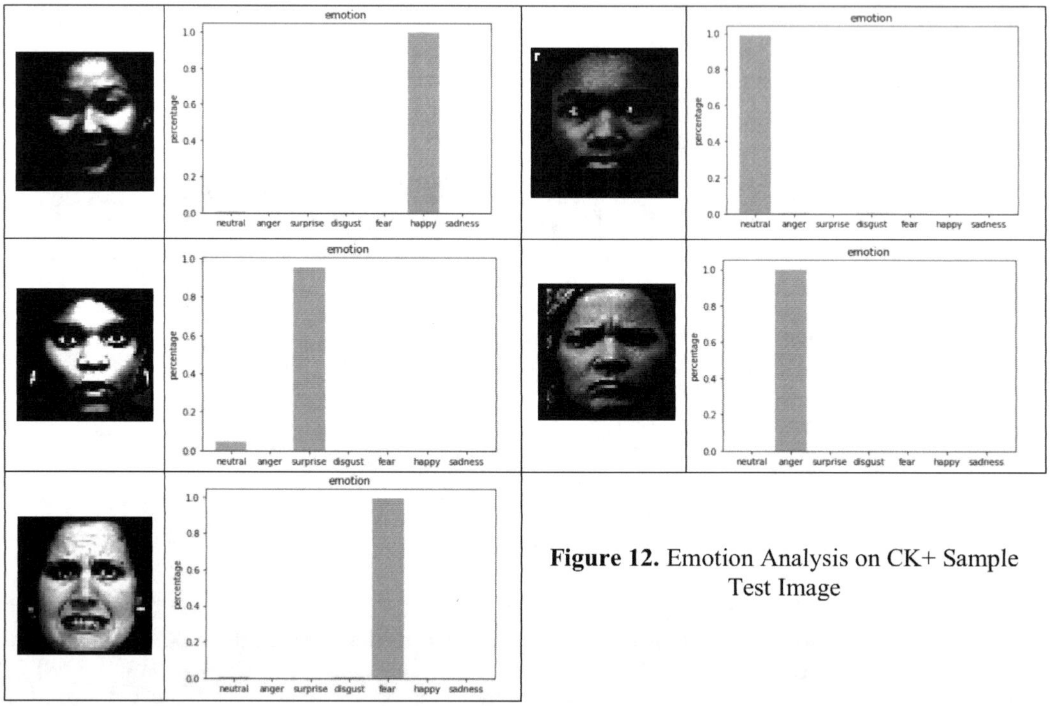

Figure 12. Emotion Analysis on CK+ Sample Test Image

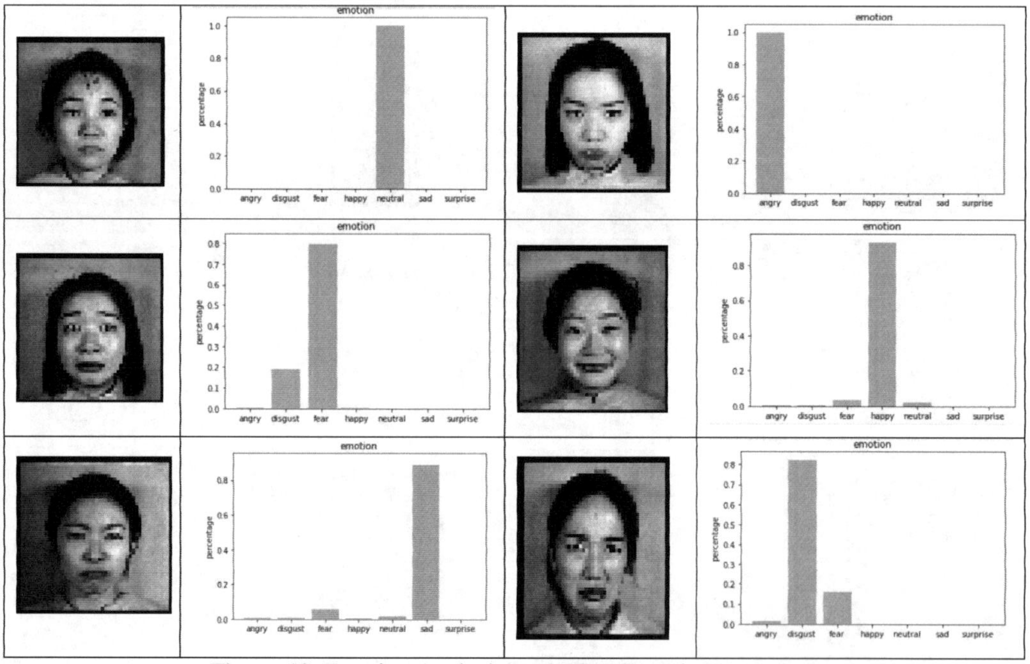

Figure 13. Emotion Analysis on JAFFE Sample Test Image

To summarize the results of the proposed methodology, refer to Table II. Table II indicates the result using the K-Fold Cross-validation technique on FER13, CK+, and JAFFE Datasets with CNN Custom Model with six layers and 1.48M parameters.

For Pre-processing, Viola-Jones (V&J) Haar cascade-based detection of frontal faces is used, and to balance training images for unbiased results, Image Data Generator is used.

Table 2. Summary results using K-Fold Cross-validation technique on FER13, CK+ and JAFFE Datasets with CNN Custom Model

Datasets	Performance(%)	K-Fold Results
FER 2013	86.78%	acc: 88.76% acc: 89.07% acc: 89.26% acc: 90.15% acc: 88.51% acc: 89.41% acc: 89.47% acc: 90.09% acc: 89.41% acc: 83.70% 86.78% (+/- 6.71%)
CK+	92.27%	acc: 93.88% acc: 94.22% acc: 92.69% acc: 93.20% acc: 92.01% acc: 94.39% acc: 94.55% acc: 93.70% acc: 93.02% acc: 81.09% 92.27% (+/- 3.80%)
JAFFE	91.58%	acc: 95.45% acc: 86.36% acc: 86.36% acc: 100.00% acc: 85.71% acc: 90.48% acc: 95.24% acc: 80.95% acc: 100.00% acc: 95.24% 91.58% (+/- 6.22%)

The model is taught seven fundamental emotion classes {"Angry," "Disgust," "Fear," "Happy," "Neutral," "Sad," and "Surprise"} from scratch. For example, the JAFFE dataset contains face photos or videos with class labels representing the fundamental emotional state of the subject. In order to get the model ready for use, Training, Validation, and Test Data are three distinct groups of facial images in a dataset.

Data sets for training and validation are split into 10 folds, and each round, a group of data Selected for training, is selected for training and accuracy calculated in each fold, as shown in K-Fold Results for each fold. Once training is complete, the trained model is utilized to determine the test dataset's Accuracy Performance (percent).

Table 3 summarizes the most recent state-of-the-art findings based on a survey [3]. Despite the fact that our custom CNN architecture has a smaller network size in terms of depth and parameter, the accuracy achieved is comparable to state-of-the-art techniques, which is remarkable.

For Table 3 abbreviations
IN = Illumination Normalization; NE = Network Ensemble; CN = Cascaded Network;
MN = Multitask Network; LOSO = Leave-One-Subject-Out: SDM= Supervised Descent Method;
AAM= Active Appearance Models; DRMF= Discriminative Response Map Fitting
RBM= Restricted Boltzmann Machine
Anger, contempt, disgust, fear, happiness, and sadness are all included in CK+'s seven classes.
Using the confusion matrix, the number in parentheses represents the mean accuracy.

Table 3. Comparison with State-of-the-art results

Datasets	Method	Network type	Network size Depth (Parameters)	Preprocessing	Data selection		Data group	Classifier	Performance (%)
FER13	Tang	CNN loss layer	4 (12M)	Yes	Training Set: 28,709 Validation Set: 3,589 Test Set: 3,589			--------	Test: 71.2
	Devries et al.	CNN MN	4 (12M)	MoT IN				--------	Validation+Test: 67.21
	Zhang et al.	CNN MN	6 (21.3M)	SDM				--------	Test: 75.10
	Guo et al.	CNN loss layer	10 (2.6M)	SDM				k-NN	Test: 71.33
	Kim et al.	CNN NE	5 (2.4M)	IntraFace IN				--------	Test: 73.73
	Pramerdorfer et al.	CNN NE	10/16/33 1.8/1.2/5.3 (M)	IN				--------	Test:75.2
	Georgescu et al.	CNN NE	8/13/16	--------				SVM	Test:75.42
	Ours	CNN loss layer	6 (1.48M)	V&J				--------	10 folds: 86.78
CK+	Ouellet	CNN (AlexNet)	--------	V&J	the last frame	7 classes†	LOSO	SVM	94.4
	Liu et al.	CNN, RBM CN	5	V&J	the last four frames and the first frame	8 classes	10 folds	SVM	92.05 (87.67)
	Liu et al.	CNN, RBM CN	5	V&J		7 classes	10 folds	SVM	93.7
	Khorrami et al.	zero-bias CNN	4 (7m)	Yes		6 classes 8 classes	10 folds	--------	6 classes: 95.7; 8 classes: 95.1
	Ding et al.	CNN fine-tune	8 (11m)	IntraFace			10 folds	--------	6 classes: (98.6); 8 classes: (96.8)
	Zeng et al.	DAE (DSAE)	3	AAM	the last three frame	7 classes† 8 classes	LOSO	--------	7 classes†: 95.79 (93.78) 8 classes: 89.84 (86.82)
	Cai et al.	CNN loss layer	6	DRMF IN		7 classes	--------	--------	94.39 (90.66)
	Meng et al.	CNN MN	6	DRMF		7 classes	10 folds	--------	95.37 (95.51)
	Liu et al.	CNN loss layer	11	IntraFace IN		7 classes†	8 folds	--------	97.1 (96.1)
	Zhang et al.	CNN MN		Yes		6 classes	10 folds	--------	98.9
	Ours	CNN loss layer	6 (1.48M)	V&J		7 classes†	10 folds	--------	92.27
JAFFE	Liu et al.	DBN	6	Yes	213 Images (7 Classes)		LOSO	AdaBoost	91.8
	Hamester et al.	CNN, CAE	3 (2M)	IN			LOSO	--------	95.8
	Ours	CNN loss layer	6 (1.48M)	V&J			10 folds	--------	91.58

6. Conclusion and Future Work

Using a custom CNN architecture, this research proposes a deep learning method for facial emotion identification. Custom CNN architecture is used to generalize the FER system for better performance on a given database-based.

To generalize the CNNs-based model for experimenting, the same structure is selected and trained independently for each dataset. For our experiments on the datasets of FER13, JAFFE, and CK+, we used a basic Custom CNN model.

Test accuracy indicates, each model was trained using the k-fold cross-validation method. During the test phase, the test sample's emotion is predicted and compared with the true label. Table 2 summarises the research findings. FER13 dataset, we are getting overall accuracy of 86.78%, which is a little less compared with CK+ accuracy 92.27% and JAFFE accuracy of 91.58%. Comparing it with state-of-art results shows implemented custom CNN giving good accuracy by considering network size in terms of depth and parameter it is less complex. However, there is room for improvement of accuracy by trying standard CNN architecture, deeper CNN network, or other approaches.

In the future, unsupervised pre-training procedures from transfer learning may be employed to minimize the amount of recognition errors even further. Pre-processing and Feature Extraction techniques can be applied before training models [5],[7]. Transfer Learning can also be tried on these datasets [8] CK+ and JAFFE Dataset have very few images, and they are not Balanced, so making them balanced will give better testing results [1],[2]. FER based on Facial Action Units (AUs) has shown good results in a state of the art that can also be tried [10]. Some Advanced Models such as DBN, VAE, and GANs can be applied to improve accuracy [9],[11].

References

[1] Schmidhuber, Jürgen. "Deep learning in neural networks: An overview." Neural networks 61 (2015): 85-117.
[2] Valter, Per, Peter Lindgren, and Ramjee Prasad. "The consequences of artificial intelligence and deep learning in a world of persuasive business models." IEEE Aerospace and Electronic Systems Magazine 33.5-6 (2018): 80-88.
[3] Li, S. and Deng, W., "Deep facial expression recognition: A survey," IEEE Transactions on Affective Computing," 2020.
[4] Krizhevsky, Alex, Ilya Sutskever, and Geoffrey E. Hinton. "Imagenet classification with deep convolutional neural networks." Advances in neural information processing systems 25 (2012): 1097-1105.
[5] Simonyan, Karen, and Andrew Zisserman. "Very deep convolutional networks for large-scale image recognition." arXiv preprint arXiv:1409.1556 (2014)
[6] Pandey, R.K., Karmakar, S., Ramakrishnan, A.G. and Saha, N., 2019. Improving Facial Emotion Recognition Systems Using Gradient and Laplacian Images. arXiv preprint arXiv:1902.05411.
[7] Pan, Sinno Jialin, and Qiang Yang. "A survey on transfer learning." IEEE Transactions on knowledge and data engineering 22.10 (2010): 1345-1359.
[8] Ramalingam, S. and Garzia, F., 2018, October. Facial Expression Recognition using Transfer Learning. In 2018 International Carnahan Conference on Security Technology (ICCST) (pp. 1-5). IEEE.
[9] Caramihale, T., Popescu, D. and Ichim, L., 2018. Emotion classification using a tensorflow generative adversarial network implementation. Symmetry, 10(9), p.414.
[10] Kim, S. and Kim, H., 2019, April. Deep Explanation Model for Facial Expression Recognition Through Facial Action Coding Unit. In 2019 IEEE International Conference on Big Data and Smart Computing (BigComp) (pp. 1-4). IEEE.
[11] Minaee, S. and Abdolrashidi, A., 2019. Deep-Emotion: Facial Expression Recognition Using Attentional Convolutional Network. arXiv preprint arXiv:1902.01019.
[12] Miao, Y., Dong, H., Jaam, J. M. A., & Saddik, A. E. (2019). A deep learning system for recognizing facial expression in real-time. ACM Transactions on Multimedia Computing, Communications, and Applications (TOMM), 15(2), 1-20.
[13] Wolfram Research, "FER-2013" from the Wolfram Data Repository (2018)
[14] P. Lucey, J. F. Cohn, T. Kanade, J. Saragih, Z. Ambadar, and I. Matthews, "The Extended Cohn-Kanade Dataset (CK+): A complete dataset for action unit and emotion-specified expression," 2010 IEEE Computer Society Conference on Computer Vision and Pattern Recognition - Workshops, 2010, pp. 94-101.
[15] Lyons, Michael, Kamachi, Miyuki, & Gyoba, Jiro. (1998). The Japanese Female Facial Expression (JAFFE) Dataset [Data set]. Zenodo.
[16] Ko, Byoung C. 2018. "A Brief Review of Facial Emotion Recognition Based on Visual Information" Sensors 18, no. 2: 401.

Journal of Physics: Conference Series **2236** (2022) 012005 doi:10.1088/1742-6596/2236/1/012005

Investigation of Electric Field Profile and associated parameters with Embedded Metal Layer in Field Plate AlGaN/GaN HEMTs

Sushanta Bordoloi[a], Ashok Ray[b], Pranjal Barman[c], Gaurav Trivedi[d]

[a]Dept. of ECE, NIT Mizoram, Aizawl 796012 Mizoram, INDIA;
[b]Dept. of ECE, NERIST, Nirjuli 791109 Arunachal Pradesh, INDIA;
[c]E&ICT Academy, IIT Guwahati, Guwahati 781039 Assam, INDIA;
[d]Dept. of EEE, IIT Guwahati, Guwahati 781039 Assam, INDIA

E-mail: sushanta.ece@nitmz.ac.in

Abstract. Electric field initiates various failure mechanism in an III-N based AlGaN/GaN HEMTs. The material properties of the GaN-HEMTs under the influence of electric field activates physical mechanisms like converse piezoelectric effect (CPE), crack/defect migration, trapping, which deteriorate the electrical behaviour of the device leading to permanent failure. The drain side of the gate edge has the highest electric field and is the hub of all the reliability concerns in a GaN HEMT, which is mitigated with field plate (FP) technology. However, the FP edge is now subjected to these degradation phenomena. The present work aims at suppressing the strong electric fields at the FP gate edge using an embedded metal layer that shields the electric field from reaching the gate edge. Calibrated numerical simulations have been carried out on the proposed device structure to observe the viability in consideration. It is found that the electric field at the FP edge reduces by around 3%. Also, CPE and electron temperature reduce by 20%, and 14%, respectively. Since the proposed device structure can considerably mitigate the electric field, CPE, and electron temperature, it is expected that it will pave a path for improved and reliable devices in the future.

1. Introduction

The initial success of a modulation type doping in an AlGaN/GaN heterostructure was followed by an inceptive realization of an AlGaN/GaN high electron mobility transistor (HEMT) in 1991 [1], and 1993 [2], respectively. GaN electronics has leaped forward with extraordinary RF-power and power switching applications [3]. The superiority of GaN-based devices compared to Si devices originates in the inherent properties of GaN material and AlGaN/GaN heterostructures. The enhanced drift velocity and higher breakdown field of GaN devices result in lower losses when the devices are operated at high voltage and temperature [4, 5]. The existence of two-dimensional electron gas (2DEG) at the interface of AlGaN/GaN in heterojunction field-effect transistor (HFET), also referred to as high electron mobility transistor (HEMT), presents immense potential for RF and power switching application [6].

Given that the operational regime of the HEMT devices belongs to both high voltage and high current, it is indispensable to comprehend the electric field profile of the device. The exceptionally high electric field at the drain side of the gate edge is a crucial factor leading

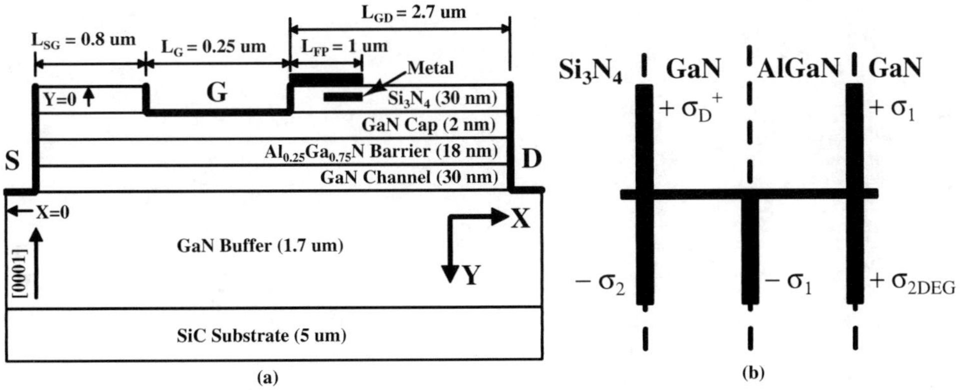

Figure 1. (a) Depiction of the proposed device structure (Not to scale). The metal plate is embedded within the passivation layer below the field plate; (b) Representation of polarization charge at the interface employed in the numerical analysis.

to early breakdown and localizing hotspots in AlGaN/GaN HEMTs. Moreover, as the device experiences a high electric field, the tensile stress increases with elastic energy in the AlGaN barrier layer. Beyond a critical value, the accumulated elastic energy eases up with the formation of crystallographic defects; this phenomenon is recognized as the converse piezoelectric effect (CPE) [7]. This adversely affects the device's performance, lifetime, and reliability, limiting the operational capability. Prior work has demonstrated the use of high-k dielectric passivation layer [8, 9], field plates (FP) [10, 11, 12, 13, 14], and modified gate geometry [15, 16, 17] to moderate the electric field at the drain side of the gate edge of the device thereby enhancing the breakdown voltage. Zhang et al. [18] demonstrated a recessed float field plate in millimeter-wave AlGaN/GaN HEMTs that modulates the electric field distribution at the drain end of the gate edge, thereby increasing the breakdown voltage.

In the proposed work, we undertake an investigative study to explore the electric field profile with an embedded metal layer within the passivation layer as the layer's position is varied. Besides, the related parameters such as converse piezoelectric strain and electron temperature are also studied. As FP can be either gate connected [10, 11, 12, 19], source connected [13], drain connected [14] or recessed float type [18], in the proposed work, we have consider the gate connected FP.

The device structure is discussed in Section 2, calibration of numerical analysis framework in Section 3, and electric field profile, converse piezoelectric effect, and electron temperature of the proposed device with embedded metal layer within passivation layer are discussed in Section 4.

2. Device Structure

The proposed idea is studied with a device structure which is illustrated in Fig. 1, along with the polarization charge profile. The device dimensions are: 0.25 μm, 2.7 μm, and 0.8 μm characterized as gate length (L_G), gate-to-drain spacing (L_{GD}), source-to-gate spacing (L_{SG}), respectively. Also, the device comprises of 30 nm Si_3N_4 passivation layer, 2 nm GaN cap layer, 18 nm $Al_{0.25}Ga_{0.75}N$ barrier layer, 30 nm GaN channel, and 1.7 μm GaN buffer on top of 5 μm SiC substrate. Silicon Carbide (SiC) is considered substrate due to better lattice match with GaN than Silicon and Sapphire. The rationale behind choosing SiC as substrate is the high thermal conductivity, which helps in better stability at higher temperatures. The field plate in the device structure is of length (L_{FP}) 1 μm.

The proposed device structure is examined with a metal plate embedded within Si_3N_4

Table 1. Material parameters used in numerical simulation

S. No.	Property of Material	Unit	GaN	AlGaN
1	Electron Mobility	cm^2/Vs	1200	300
2	Bandgap	eV	3.50	4.39
3	Electron Affinity	eV	4.0	3.41
4	Relative Permittivity	-	8.9	8.8
5	Electron Saturation Velocity	cm/s	2.5×10^7	1.1×10^7
6	Electron Conduction Band Density of States (DOS)	cm^{-3}	2.23×10^{18}	2.71×10^{18}
7	Electron Valence Band Density of States (DOS)	cm^{-3}	2.51×10^{19}	2.06×10^{19}

passivation layer below the field plate at the edge. For the numerical analysis, the dimension of the embedded metal layer is; 5/10 nm thick and 100 nm in length. The depth/vertical distance from the field plate is considered in the step of 5 nm till the metal layer reaches the boundary of the passivation layer. The objective is to estimate the optimum location of the metal layer within the passivation layer.

3. Numerical Simulation Framework: Basics and its Calibration

The device structure proposed in this work is numerically analyzed after designed using Synopsys Sentaurus TCAD Tool [20]. The model included in the numerical simulation are; Poisson and continuity equation for both holes and electrons, drift diffusion model, and Shockley-Read-Hall recombination model. Constant as well as field-dependent mobility for carriers are included. The recombination and generation of carriers are accounted by Shockley-Read-Hall model. The development of 2DEG at the interface of AlGaN/GaN due to piezoelectric and spontaneous polarization, eliminates the prerequisite of doping in the AlGaN barrier layer. The polarization charges considered for the numerical simulation is shown in Fig. 1(b); $\sigma_1 = \pm\ 1.23 \times 10^{13}$ cm^{-2}, and σ_2 = - 2.0×10^{12} cm^{-2}. Beside, donor traps (σ_D^+) with a density of 1.0×10^{13} cm^{-2} and energy level of 0.2 eV below the conduction band is introduced at the Si_3N_4/GaN interface. The buffer is doped by using having a concentration of 3×10^{16} cm^{-3} and activation energy, E_{AC} = 0.9 eV above GaN's valence band. The acceptor traps having a concentration density of 5×10^{16} cm^{-3} and an energy level of 0.4 eV below the conduction band are considered. The gate contact is Schottky with a work function of 5.2 eV. The material properties implemented in numerical simulation are stated in Table 1.

The proposed device without embedded metal in the SiN passivation layer resembles the device investigated by [21]. The numerical analysis framework considering the scenario of the proposed device without embedded metal in the SiN passivation layer is calibrated with that of experimental data of [21]. The device characteristics; output (I_{DS} versus V_{GS}), and transfer (I_{DS} versus V_{DS}) of the numerical analysis framework and experimental results are exhibited in the Fig. 2. It is evident that both the characteristics, i.e., one from numerical simulation and other from experimental data are in almost similar, which reaffirms the proper selection of the device parameters and physical models.

4. Results & Discussion

In this section, the result of the proposed device is stated. The length of the embedded metal layer is fixed as 100 nm. The thickness of the embedded metal layer is considered 5 nm and 10 nm, respectively. Likewise, the position of the embedded metal layer is varied with the SiN passivation layer. For the 5 nm thickness metal layer, the position within the passivation layer is denoted as T1, T2, T3, T4, and T5. Similarly, for the 10 nm thickness metal layer, the position

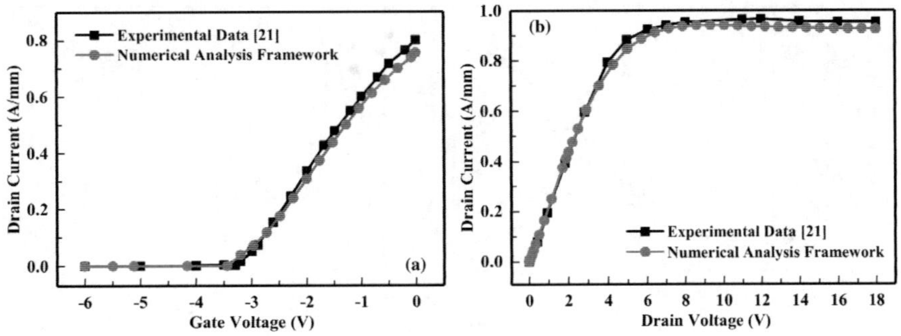

Figure 2. Comparison of DC characteristics extracted from the numerical analysis framework with that of experimental data of [21]: (a) I_D versus V_{GS}; (b) I_D versus V_{DS}

Table 2. Notation of position for metal layer within SiN passivation layer

Thickness of Metal Layer	Notation	Position (Y-Axis)
5 nm	T1	(0.005)
	T2	(0.010)
	T3	(0.015)
	T4	(0.020)
	T5	(0.025)
10 nm	P1	(0.005)
	P2	(0.010)
	P3	(0.015)
	P4	(0.020)

within the passivation layer is denoted as P1, P2, P3, and P4; refer to Table 2 for details. In the subsequent text, an embedded metal layer with 5/10 nm thickness is referred to as a thin/thick embedded metal layer.

4.1. DC Characteristics

In this section, the DC characteristics of the devices with different thickness of metal and their position within the passivation layer is illustrated.

The transfer characteristics (I_{DS} versus V_{GS}) and transconductance (g_m) for the devices are shown in Fig 3. It is evident from Fig. 3 (a) and (c) that for thicker metal within the passivation layer magnitude of decrease in drain current is more. This is attributed to the shielding effect of the electric field by the embedded metal layer below the field plate within the passivation layer. The shielding effect increases as the thickness of the embedded metal layer increases. Besides, it is observed that peak transconductance is more in the device with thinner embedded metal as compared to a device with thicker metal (refer Fig. 3 (b) and (d)). It can be concluded that devices with thinner embedded metal exhibit transconductance nearly the same as a device without a metal layer. At the same time, degradation of transconductance is observable in devices with thicker embedded metal.

Similarly, the output characteristics (I_D versus V_{DS}) of the proposed devices are shown in Fig. 4. The output characteristics of the devices with thin and thicker embedded metal layers follow a similar trend of the transconductance and transfer characteristics. The drain current of the devices with thinner metal hardly changes, while devices with thicker metal reduction are

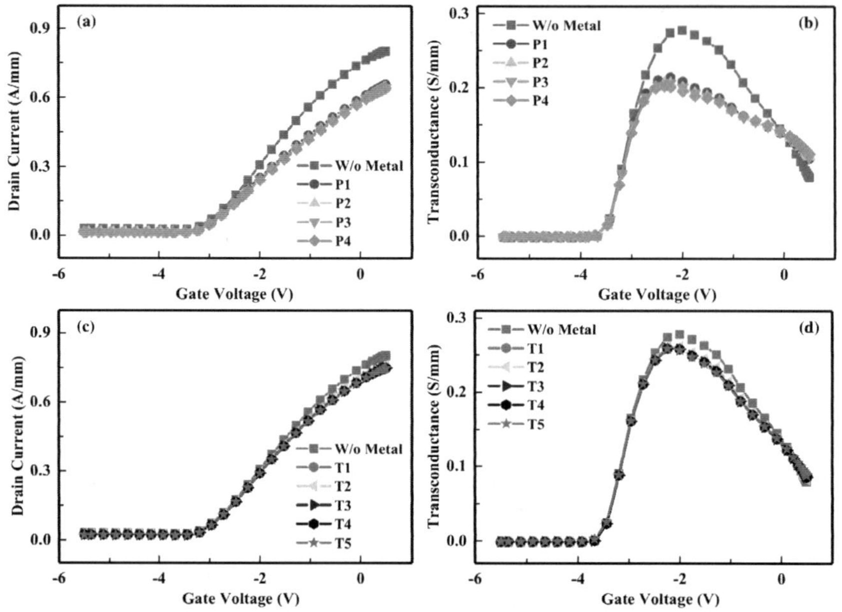

Figure 3. Comparison of DC characteristics, namely I_D versus V_{GS}, g_m of the proposed device; (a) and (b) for thick metal layer, and (c) and (d) for thinner metal layer, respectively.

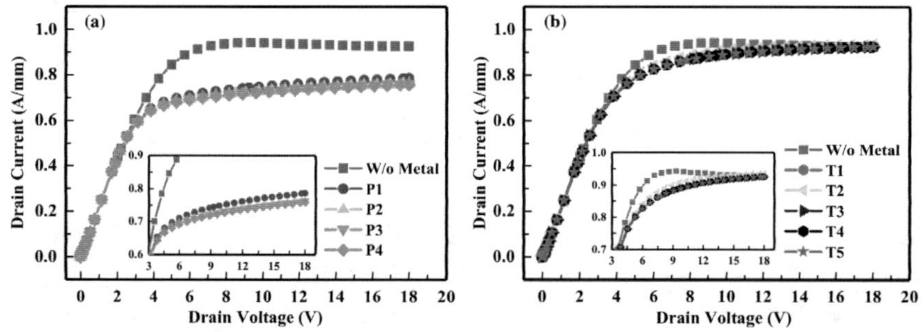

Figure 4. Comparison of DC characteristics, I_D versus V_{DS}; for (a) thicker embedded metal layer, and (b) thinner embedded metal layer.

noticeable.

4.2. Electric Field Profile

The implementation of FP technology moderates electric field crowding at drain end of the gate edge. Even, with incorporation of FP in device, it is observed that an extra electric field peak is visible at the edge of the FP. As there is less distance between the edge of FP and drain contact of the device, the high electric field at FP edge can trigger localized Schottky-barrier breakdown [22], along with thermally-assisted tunneling and avalanche breakdown in the device [23]. So, we propose to utilize an embedded metal layer to moderate this high electric field at the edge of FP.

The vertical electric field profile of the proposed devices is depicted in Fig. 5. It is observed that with the introduction of the metal layer beneath the field plate, an additional peak appears at the left edge of the metal layer (refer P1, P2, P3, and P4 of Fig. 5(a)). As the right edge

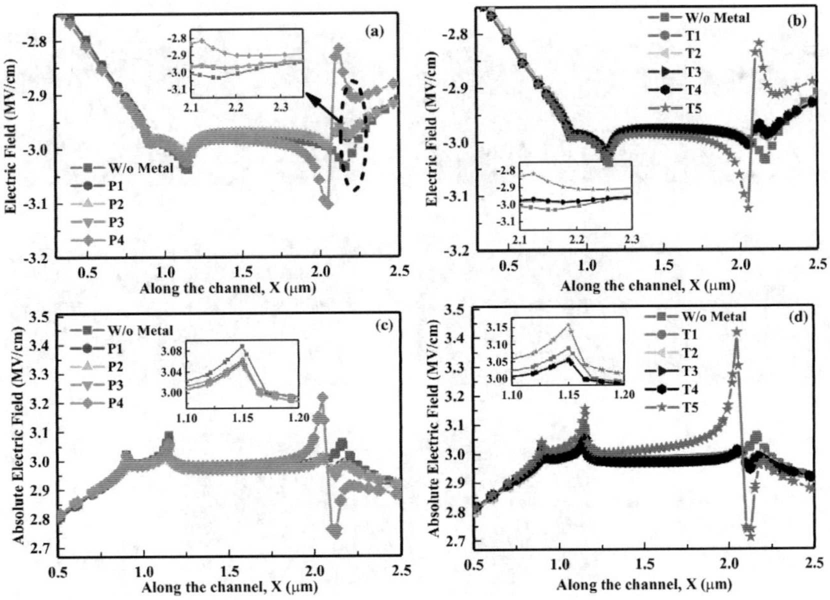

Figure 5. Comparison of electric field characteristics for the metal plate position within SiN passivation layer; for (a) thicker embedded metal layer, and (b) thinner embedded metal layer

of the metal layer is vertically aligned with the edge of the FP, a single peak is observed at the right edge of the embedded metal layer. The peak electric field at the left edge of the embedded metal layer is less than that of the FP edge when the device does not have an embedded metal layer. This is true for the scenario when the metal layer is positioned within the passivation layer (refer P1, P2, & P3). When the metal layer touches the Si_3N_4 layer, the electric field shoots up at the left edge of the embedded metal layer (refer P4 of Fig. 5(a)). Further, the electric field peak at the edge of FP of the device with embedded metal layer decreases as compared to the peak at the FP edge of the device without a metal layer. This observation is similar even when the thickness of the metal layer is decreased (refer T1, T2, T3, T4, & T5 of Fig. 5(b)).

The detailed comparison of peak electric field at the edge of the metal layer is stated in Table 3. Thus, the peak electric field is minimum for a device with metal positioned at mid-level and rises rapidly as the embedded metal layer touches the boundary of the passivation layer. Hence, the optimum position for the embedded metal layer is T3 and P2 in the case of a thinner and thicker embedded metal layer. Additionally, the absolute electric field of the devices with thick and thin embedded metal layers are illustrated in Fig. 5(c), and 5(d), respectively.

4.3. Converse Piezoelectric Profile

It is well known that stress is inherently present in the AlGaN/GaN heterostructure due to its lattice mismatch. Besides, it is well studied that the piezoelectric nature of AlGaN/GaN heterostructure is responsible for triggering the Converse Piezoelectric (CPE) phenomena. This phenomenon generates additional stress, which finally adds to the existing stress. This stress is accumulated in the form of elastic energy, and beyond a limit, it starts to generate cracks and crystallographic issues in the device [24].

For CPE to kick in, an external electric field is mandatory. Hence, reducing the electric field at hotspots is crucial for enhancing the device's longevity. The cracks evolve mainly due to excessive strain, which relaxes after the creation of defects [7, 24]. The migration of cracks in a device is diffusive in nature as it follows the square-law dependence [25]. The electric field in

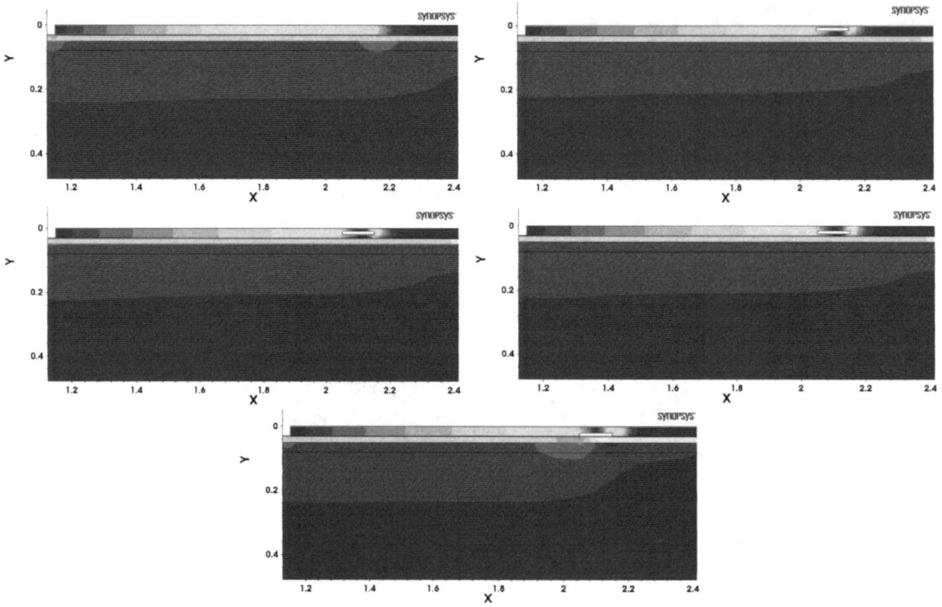

Figure 6. Depiction of E-Field profile of the thick embedded metal layer as its position varies within SiN passivation layer

Table 3. Peak Electric field at the left edge of metal layer [at X = 2.05 μm], and at the right edge of metal layer (or at field plate edge) [at X = 2.15 μm] as the position of metal layer is varied with the passivation layer. The change (in %) for peak electric field at the right edge of metal layer as the position of metal layer is varied with respect to the peak electric field at the edge of field plate for device without metal layer.

Devices	E-Field (MV/cm) at left edge of metal layer, % Change	E-Field (MV/cm) at right edge of metal layer, % Change
W/o Metal Layer	-3.001	-3.039
T1	-2.999 (↓ 0.07%)	-2.987 (↓ 1.71%)
T2	-2.998 (↓ 0.10%)	-2.985 (↓ 1.78%)
T3	-2.998 (↓ 0.10%)	-2.981 (↓ 1.91%)
T4	-3.000 (↓ 0.03%)	-2.978 (↓ 2.01%)
T5	**-3.124** (↑ 4.1%)	-2.865 (↓ 5.72%)
W/o Metal Layer	-3.001	-3.039
P1	-2.995 (↓ 0.19%)	-2.974 (↓ 2.14%)
P2	-2.994 (↓ 0.23%)	-2.970 (↓ 2.27%)
P3	-2.995 (↓ 0.19%)	-2.967 (↓ 2.37%)
P4	**-3.103** (↑ 3.39%)	-2.859 (↓ 5.92%)

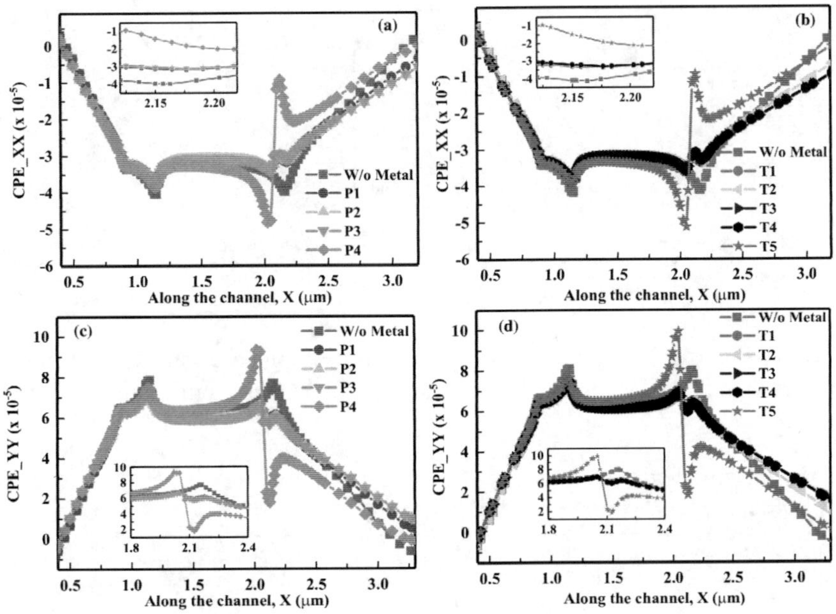

Figure 7. Comparison of planar CPE, namely, CPE_XX, and CPE_YY for the various metal plate position within SiN passivation layer: (a), & (c) for thicker embedded metal layer, and (b) & (d) for thinner embedded metal layer

FP HEMTs peaks at two locations: at the drain side of the gate edge and the end of the field plate. Hence, the proposed embedded metal layer is positioned so that the edge of FP aligns with the right edge of the metal layer. It helps shield the coupling of the electric field peak to the layers underneath it. The numerical analysis is performed for different positions of the embedded metal layer within the passivation layer in order to determine the optimum position. In Fig. 6, it is clearly observed that the embedded layer shields the electric field at the FP edge and its effect on the underlying layer. In Fig. 7, planar CPE strains, namely, CPE_XX and CPE_YY of the devices with different positions of the embedded metal layer, is illustrated.

4.4. Electron Temperature profile

Hot electrons or electrons with higher temperatures increase trap density under the gate and gate-drain access region [26]. The electron energy is reported to be one of the contributing factors for the negative differential output conductance (NDC) in a device, regardless of its operational power. Besides, NDC is due to a hot electron being captured in bulk traps [26, 27]. This indicates that electron temperature plays a far more crucial role in device degradation than lattice temperature. Moreover, the localized strain and/or electron temperature further accelerate the damage due to electron injection. Thus, it is expected that the combined effect of damages due to elevated temperatures and/or electron injection weakens the barrier layer. Hence, it is essential to examine the electron temperature profile of the proposed devices.

The electron temperature (e-Temp) profile of the devices is shown in Fig. 8. The peak e-Temp of the devices without embedded metal layer, P1, P2, P3, and P4, are 563 K, 493 K, 489 K, 490 K, and 541 K, respectively. The peak e-Temp decreases with the metal layer being embedded beneath the field plate, which is valid for all scenarios when the metal layer is within the passivation layer. This observation is true for the device with a thin embedded metal layer; T1, T2, T3, and T4.

When the metal layer touches the passivation layer boundary, e-Temp increases compared

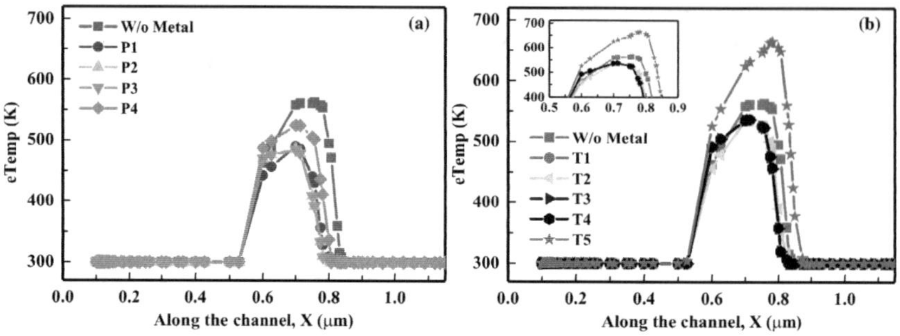

Figure 8. Comparison of e-Temp profile for the metal plate position within SiN passivation layer; (a) for thicker embedded metal layer, and (b) for thinner embedded metal layer

Table 4. Peak e-Temp for the devices, along with change (in %) with respect to maximum e-Temp for the device without metal layer.

Devices	Peak e-Temp (K), % Change
W/o Metal Layer	563
T1	536 (↓ 4.79%)
T2	535 (↓ 4.97%)
T3	534 (↓ 5.15%)
T4	536 (↓ 4.79%)
T5	**674** (↑ 19.72%)
W/o Metal Layer	563
P1	493 (↓ 12.43%)
P2	489 (↓ 13.14%)
P3	490 (↓ 12.97%)
P4	541 (↓ 3.91%)

to the scenario when the metal layer is positioned within the passivation layer due to increased coupling of the electric field; for instance, e-Temp of P4 is more than that of P1, P2, & P3. Likewise, the e-Temp of T5 is substantial compared to T1, T2, T3, and T4. This observation of e-Temp with change in position of the metal layer can be attributed to the pattern of electric field as described in the preceding section 4.2. The detailed e-Temp of the devices with the embedded metal layer is stated in Table 4.

5. Conclusion

A systematic study of the electric field, CPE, and e-Temp of an FP AlGaN/GaN HEMT with an embedded metal layer is presented. It is observed that electric field, CPE, and e-Temp are minimum when the metal layer is positioned at the center of the passivation layer. Whereas when the embedded metal layer touches the boundary of the passivation layer, electric field, CPE, e-Temp shoots up. This is valid for both thin and thick embedded metal layers. Hence, better mitigation of electric field, converse piezoelectric effect, and electron temperature is expected for the embedded metal layer to be positioned at the passivation layer's center. The proposed approach can be an alternative method to deal with race in the electric field and its associated parameter at the FP edge of AlGaN/GaN HEMTs.

References

[1] Khan, M. A., et al.: 'High electron mobility GaN/$Al_xGa_{1-x}N$ heterostructures grown by low-pressure metal organic chemical vapor deposition', *Appl. Phys. Lett.*, 1991, **58**, (21), pp. 2408-2410.
[2] Khan, M. A., et al.: 'High electron mobility transistor based on a GaN-$Al_xGa_{1-x}N$ heterojunction', *Appl. Phys. Lett.*, 1993, **63**, (9), pp. 1214-1215.
[3] He J., et al.: 'Recent Advances in GaN-Based Power HEMT Devices', *Advanced Electronic Materials*, 2021, **7**, pp. 2001045
[4] Quay R.: 'Gallium Nitride Electronics', *Springer Science & Business Media*, 2008, **96**, New York
[5] Meneghini, M., et al.: 'Power GaN Device', *Springer International Publishing*, 2017.
[6] Wosko, M., et al.: 'MOVPE growth conditions optimization for AlGaN/GaN/Si heterostructures with SiN and LT-AlN interlayers designed for HEMT applications', *J. Mater. Sci.: Mater. Electro.*, 2019, **30**, (4), pp. 4111–4116.
[7] Joh, J., et al.: 'Critical voltage for electrical degradation of GaN high-electron mobility transistors', *IEEE Electron Device Lett.*, 2008, **29**, (4), pp. 287-289.
[8] Kabemura, T., et al.: 'Enhancement of breakdown voltage in AlGaN/GaN HEMTs: field plate plus high-k passivation layer and high acceptor density in buffer layer', *IEEE Trans. Electron Dev.*, 2018, **65**(9), pp. 3848-3854.
[9] Nakamura, K., et al.: 'Analysis of breakdown voltages in AlGaN/ GaNHEMTs with low-k/high-k double passivation layers', *IEEE Trans. Device Mater. Rel*, 2019, **19**(2), pp. 298-303.
[10] Karmalkar, S., et al.: 'Enhancement of breakdown voltage in AlGaN/GaN high electron mobility transistors using a field plate', *IEEE Trans. Electron Devices*, 2001, **48**, (8), pp. 1515-1521.
[11] Onodera, H.: 'Analysis of buffer-impurity and field-plate effects on break-down characteristics in small sized AlGaN/GaN high electron mobility transistors', *Semicond. Sci. Technol.*, 2012, **27**(8), pp. 085016-1-6.
[12] Prasannanjaneyulu, B., et al.: 'Relative effectiveness of high-k passivation and gate-connected field plate techniques in enhancing GaN HEMT breakdown', *Microelectronics Rel.*, 2020, **110**, p. 113698.
[13] Xie, G., et al.: 'Breakdown-voltage-enhancement technique for RF-based AlGaN/GaN HEMTs with a source-connected air-bridge field plate', *IEEE Electron Device Lett.*, 2012, **33**(5), pp. 670-672.
[14] Soni, A., et al.: 'Novel Drain-Connected Field Plate GaN HEMT Designs for Improved V_{BD} - R_{ON} Tradeoff and RF PA Performance', *IEEE Trans. Electron Devices*, 2020, **67**, (4), pp. 1718-1725.
[15] Bordoloi S., et al.: 'Introspection Into Reliability Aspects in AlGaN/GaN HEMTs With Gate Geometry Modification', *IEEE Access*, 2021, **9**, pp. 99828–99841.
[16] Ray A., et al.: 'Numerical Simulation of Enhanced-Reliability Filleted-Gate AlGaN/GaN HEMT', *J. Elec Mater.*, 2020, **49**, pp. 2018–2031.
[17] Bordoloi, S., et al.: 'Numerical analysis of the Impact of Gate Geometry variations on the Reliability of AlGaN/GaN HEMT', *IEEE 4th International Conference on Computing, Power and Communication Technologies (GUCON)*, 2021, pp. 1-5.
[18] Zhang, S., et al.: 'Millimeter-wave AlGaN/GaN HEMTs Breakdown Voltage Enhancement by a Recessed Float Field Plate', *Appl. Phys. Express*, 2019.
[19] Bordoloi, S., et al.: 'Access Region Stack Engineering for Mitigation of Degradation in AlGaN/GaN HEMTs with Field Plate', *IEEE Trans. Device Mater. Rel.*, 2022, doi:10.1109/TDMR.2022.3150714.
[20] Synopsys: 'Sentaurus Device User Guide', *Version L-2016.03 ed.*, 2016.
[21] Subramani, N. K., et al.: 'Identification of GaN Buffer Traps in Microwave Power AlGaN/GaN HEMTs Through Low Frequency S-Parameters Measurements and TCAD-Based Physical Device Simulations', *IEEE J. Electron Devices Society*, 2017, **5**, (3), pp. 175–181.
[22] Xing, H., et al.: 'High breakdown voltage AlGaN-GaN HEMTs achieved by multiple field plates', *IEEE Electron Dev. Lett.*, 2004, **25**, (4), pp. 161-163.
[23] Trew, R. J., et al.: 'Gate breakdown in MESFETs and HEMTs', *IEEE Electron Dev. Lett.*, 1991, **12**, (10), pp. 524-526.
[24] del Alamo, J. A., et al.: 'GaN HEMT Reliability', *Microelectronics Rel.*, 2009, **49**, (9-11), pp. 1200-1206.
[25] Ancona, M.G., et al.: 'Fully coupled thermoelectromechanical analysis of GaN high electron mobility transistor degradation', *J. Appl. Phys.*, 2012, **111**, (7), p. 074504.
[26] Tapajna, M., et al.: 'Hot-electron-related degradation in InAlN/GaN high-electron-mobility transistors', *IEEE Trans. Electron Dev.*, 2014, **61**, (8), pp. 2793-2801.
[27] Braga, N., et al.: 'Simulation of hot electron and quantum effects in AlGaN/GaN heterostructure field effect transistors', *J. Appl. Phys.*, 2004, **95**, (11), pp. 6409-6413.

A sustainable embedded shade management system for tea cultivation

Hemarjit Ningombam[1*], O.P Roy[2]

[1]Department of Electronics & Communication Engineering, Dibrugarh University, Assam, India

[2]Department of Electrical Engineering, NERIST, Nirjuli, Arunachal Pradesh, India

*Corresponding author's e-mail address: nhsingh@dibru.ac.in

Abstract: Tea farming stands on a vast area. The quality and quantity of products enormously depend on the management and control of the environmental conditions of the tea farms. Feasible and continuous monitoring of the plantations' parameters is needed to achieve the desired farm output. 50%-60% of the light intensity needs to be filtered on each tea plant throughout the year. Traditionally, shading is achieved by planting leguminous trees inside the farms. So it is extensively required to monitor the shading capacity continuously to overcome the degradation of the farm's product. In this paper, a simple, sustainable shade management system is designed. The system is embedded with Atmega 2560 controller, and a wireless network has been achieved using the ZigBee module. A systematic array of LDRs acquires light Intensity. The measured parameters are wirelessly transmitted continuously to the central decision support system with associated information like locations, sensor values, etc. The proposed method is easily extensible and user-friendly.

Keywords: Embedded sensing module, Zigbee wireless, remote decision support, shade management, tea cultivation

1. Introduction:

It is believed that tea thrives well under shade. In Assam, tea is grown under the shade trees. Generally, all the tea cultivation in the Assam sau tree (albizzia chinesis) was planted in tea plantations and considered as tea fertilizer tree. Shade tree not only provides nutrients to the soil, but it also has many benefits to the tea plants in those areas where temperature rises to affect photosynthesis and cause damage to leaves by scorching[1][2].

Environmental temperature below 13^0 C and above 30^0C is not suitable for the growth of tea. However, in northeast India, mainly Assam, the air temperature lies above 30^0C for a more significant part of the harvesting season. Hence, shade trees are interplanted in tea farms to lower the temperature below a level where it cannot adversely affect the photosynthesis process [3].

Regions where wind velocity is high, like Darjiling, helps in reducing leaf temperature and do not require shade trees. But wind velocity in Assam is low due to the shelter effect of hills on either side of the Brahmaputra valley. For that, tea is cultivated under the shade trees in the region [4].

1.1 Need of Shade Management:

Besides regulating the ambient temperature improving soil fertility through enhanced microbial and earthworm activity, shade trees have other ecological influences on tea diseases and pests. Outbreaks of disease like Blister blight is enhanced under high humid environmental conditions. Therefore, in locations or regions where tea is prone to this disease, it is necessary to regulate the shade [1]. High humidity helps control pests such as mites and thrips as a natural biological control mechanism of these pests, and it is achievable only under humid environmental conditions [1][4]. An optimum number of shade trees per hectare be retained in the tea estates, allowing the required sunshine (50-70%) to break through the canopy for optimum growth of tea plants. Lopping and thinning shade trees should be done immediately, following the proper procedure when the shade is heavy[4][1]. Care should be taken so that the tea crop is not damaged much. Sometimes proper spacing of shade trees after thinning is not maintained. Some trees die, but the gap is not filled again. As a result, an even canopy of shade trees is not found over the tea plantation. In some patches, tea plants get a lower amount of sunlight than necessary, while others are flooded with the sun, which is detrimental to tea growth[1][5]. The traditional way of monitoring and managing the vast coverage of tea plantations degrade yield quality. Therefore, accurate and timely response requires the situation. To overcome the problems faced by the tea farmers, a wireless shade management system to monitor the amount of light intensity continuously on the tea farm is proposed herewith.

2. System components:

2.1. LDR (Light Dependent Resistor)

A Light Dependent Resistor (LDR) or photoresistor is a light-sensitive device that means the resistivity of LDR is the function of the incident light on it. The LDR works on the principle of photoconductivity which is an optical phenomenon [6]. When a constant electromotive force (emf) is applied to a Light Dependent Resistor, the current also starts increasing if the incident light intensity increases. Figure 1 shows the graph of resistance as a function of illumination. Illumination decreases as the resistance of the LDR increases.

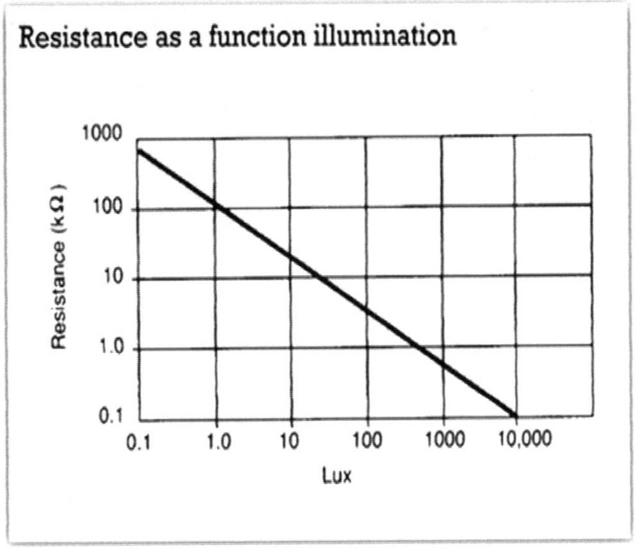

Figure 1. Resistance vs. illumination

Journal of Physics: Conference Series **2236** (2022) 012006 doi:10.1088/1742-6596/2236/1/012006

2.2. Atmega 2560 board:

It is based on Atmega 2560 microcontroller from Microchip, as shown in figure 2. Atmega 2560 is the high-performance, low-power 8-bit RISC (Reduced Instruction Set Computer)-based microcontroller that combines 256 KB ISP (In-System Programming) flash memory, 8 KB SRAM (Static RAM), 4 KB EEPROM (Electrically Erasable Programmable ROM), 86 general purpose I/O lines, 32 general purpose working registers, real-time counter, six flexible timer/counters with compare modes, PWM (Pulse Width Modulation), four USARTs (Universal synchronous and asynchronous receiver-transmitter), byte-oriented Two-Wire serial interface, 16-channel 10-bit A/D converter, and a JTAG interface for on-chip debugging. The device achieves a throughput of 16 MIPS (Million instructions per second) at 16 MHz and operates between 4.5-5.5 volts [7].

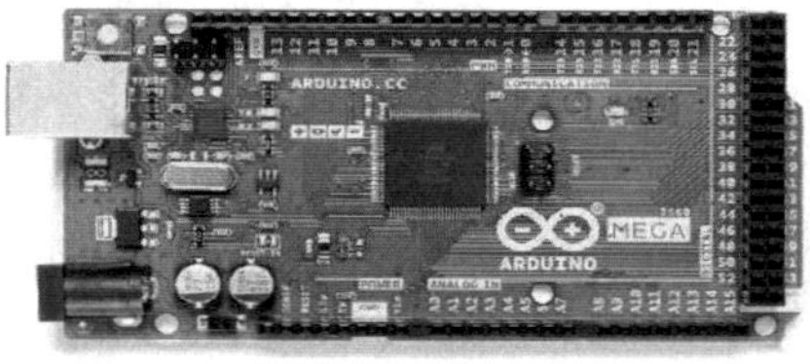

Figure 2. Atmega 2560

Figure 3. XBee RF module

2.3. ZigBee/802.15.4 RF Modem:

The XBee/XBee-PRO Zigbee RF Modules, figure 3, provide wireless connectivity to end-point devices in Zigbee mesh networks. It transmits data up to 1 mile for outdoors range (LOS) and 100m in indoor range at frequency 2.4 GHz[8].The receiver sensitivity is -100 dBm. One Xbee module can be replaced by other Zigbee devices, including devices from different vendors[8]. With the XBee, users can have their Zigbee network up and running in a matter of minutes without configuration or additional development. The XBee/XBee-PRO Zigbee RF Modules are compatible with other devices that use XBee Zigbee technology[9].

3. Sensing module and deployment:

3.1. Light intensity measurement:

As shown by the properties of LDR, the resistance of the LDR sensor decreases with light Intensity or the conductance between two contacts of sensor increases. As shown in figure 4, we convert the change in resistance to a change in voltage using a simple voltage divider circuit. In the divider circuit, R1 is constant, and R2 (LDR) varies. The amount of variation is considered the measured parameter and given by the output equation 1, taken from the midpoint of the branch.

$$Vout = (R2/(R1 + R2))Vin \text{ -------------- } (1)$$

When resistance (R2) changes to output, $Vout$ changes with it linearly. So, we obtain a voltage that is changing with the change in Intensity of light.

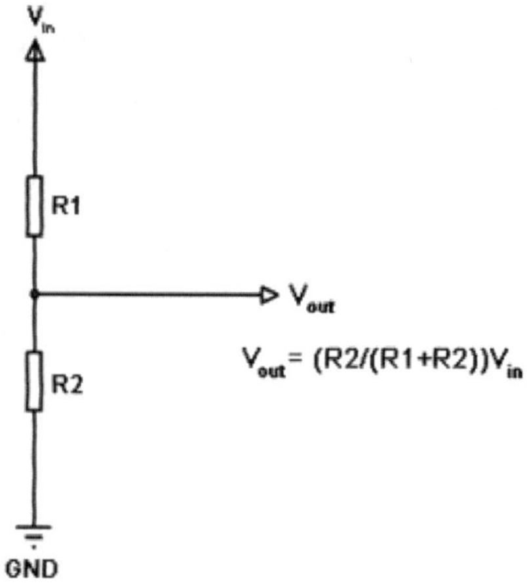

Figure 4. Voltage divider circuit

As the aim of the project is not to measure precise data, we convert the output in Lux (approximate) using a typical LDR relation as follows

$$R2 = 500/\text{Lux} \text{ .... } (2)$$

Supplying Vin of 5V through a 10K resistor, using voltage divider equation 1, the output voltage of the LDR is

$$\text{Vo} = (5 \times R2)/(R2 + 10K) \text{ .... } (3)$$

From equation 1 into equation 2,

$$\text{Lux(approximate)} = [(2500/\text{Vo}) - 500]/10K \text{ .... } (4)$$

As the output of the circuit, we find the serial monitor display as figure 5 below.

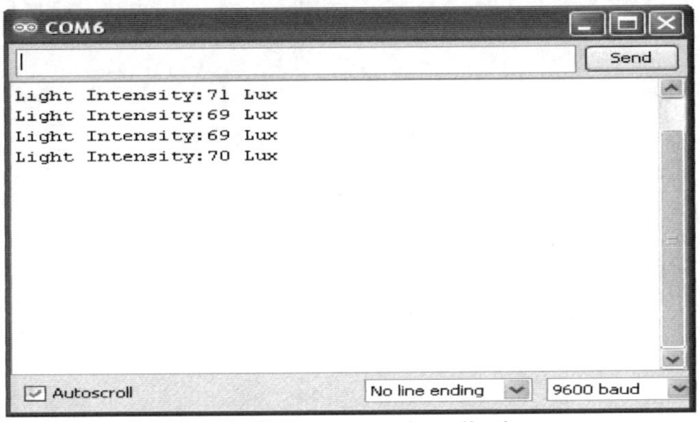

Figure 5. Serial monitor display

3.2. Experimental setup:

The surface area covered by a Light-dependent resistor is minimal and unsuitable to use in a wide surface application. In this experiment, to minimize the problem, to increase the surface contact of the sensor, uniformly distributed, eight quantities of LDR are connected in parallel with a supply voltage of 5 volts and a pull-up resistor of 10K for each sensor module.

Three or four sensor modules are connected and kept in control by each node, as shown in figure 6. The sensor modules are deployed to cover the optimum space of the farm around the nodes. Sum or average values of the sensor readings are continuously transmitted to a remote central decision support system through the wireless network. Then the central decision support system accesses the received sensors reading and proceed for further processes.

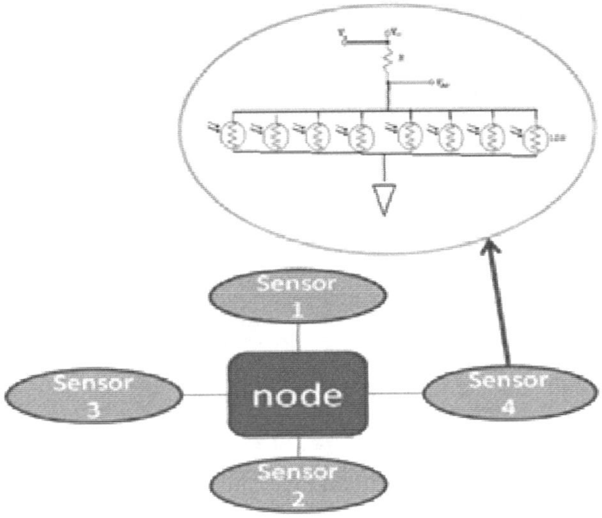

Figure 6. Experimental setup

Journal of Physics: Conference Series **2236** (2022) 012006 doi:10.1088/1742-6596/2236/1/012006

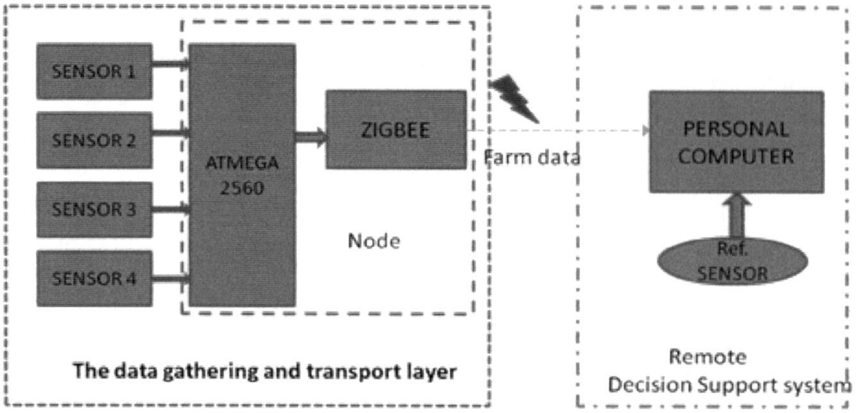

Figure 7 System block diagram

For calibration purposes, a reference sensor module is installed at the end of the decision support system, as shown in figure 7. This sensing module measures the Intensity of the uncovered farm area where there is no shade tree. The Decision support system formulates the final output using the below equation.

$$\mathbf{S(n)out = URef - U(n)Control}$$

URef is the received output of the sensor module placed outside the shaded area connected to the decision support system. **U(n)Control** is the received output of the (n) number of sensor modules located at the shaded area of the farm. **S(n)out** denote the calculated final output given by the central decision support system, whose value Should be between the range of 50 to 60% of the **URef** to satisfied the shaded condition.

4. Sensing node development:

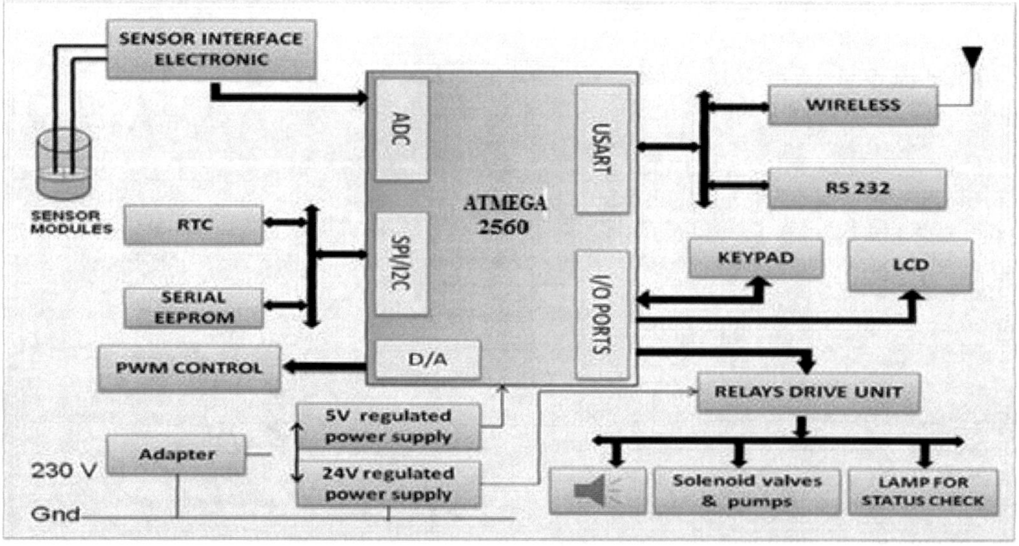

Figure 8. A wireless sensing unit

Journal of Physics: Conference Series **2236** (2022) 012006 doi:10.1088/1742-6596/2236/1/012006

A wireless sensing unit is developed using Atmega 2560 microcontroller, as shown in figure 8. It is designed to monitor and control different parameters required in the tea cultivation process. The unit mainly consists of a sensing block, control block, wired or wireless communication, and data storage block. It has the provision to control sprinkler system for tea farm irrigation. The unit is sufficiently small in size and can interface with multiple sensors.

In this context of shade management in tea cultivation, each sensing node is associated with four LDR assemblies mentioned above and four temperature sensors. These total eight sensors are connected to the analog/ADC port of the Atmega 2560 microcontroller[2]. The 10 bits with acquisition time 12.86μs onboard ADC of Atmega 2560 sampled and quantized the analog inputs from the sensors[10]. Then, using the respective equations, the controller calibrated the temperature to degree Celsius and light intensity to Lux. Moreover, the microcontroller regulates the sensors' scanning rate of the ADC, and it calculates the average of all sensor readings both for temperature and light intensity. Average sensor parameters are transmitted to the central decision support system for further computation. The parameter values are transmitted through wireless communication using the Zigbee protocol. X-bee module has transmission strength up to 1.6 Kilometers line of sight at frequency 2.4GHz[9].To avoid any data losses during real-time wireless communication, acquisition time and conversion time should be fast enough[11].

4.1 X-bee network deployment:

There are three different devices of the Zig-bee protocol; coordinator, router, and end device.

Coordinator: the coordinator act as the center of the network. It initiates a new PAN and allows router and end devices to join the network by selecting PAN and channel. The coordinator carries out both transmission and reception functions along with routing information to the network[12]. It collects sensing parameters from the sensors and transmits them back to the user or control platform.

Router: It functions as the intermediate between the sensor nodes. Router received information from the coordinator and relay to end devices and vice versa[13].

End device: The end device has sufficient function to communicate with the coordinator device and router device but cannot relay information from other devices[13]. End devices are usually sensors and are considered power-saving devices.

Generally, Zig-bee mesh topology is considered a flexible and reliable topology[12]. This topology consists of a coordinator, a few routers, and many end devices. A router can create a link with other routers and end devices. Without the relay of the coordinator, routers within the communicable range can link among each other, and hence mesh topology is treated as a flexible topology[14]. Mesh network is reliable because an alternative link is created when there is a failure on the existing link. This topology applies to both wireless networks and wired networks. A simple mesh network is shown in figure 9 below.

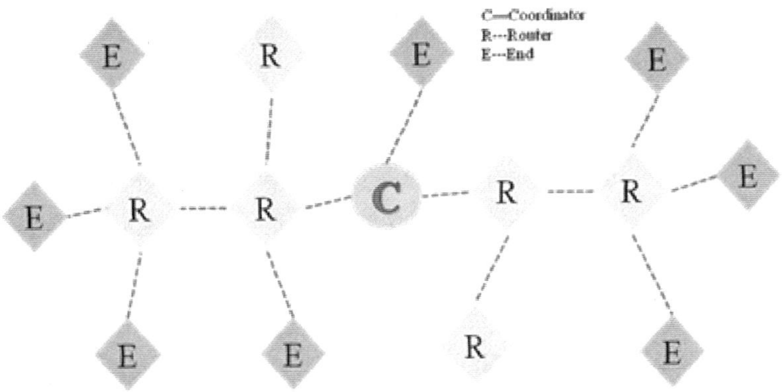

Fig. 9. Mesh topology

Journal of Physics: Conference Series **2236** (2022) 012006 doi:10.1088/1742-6596/2236/1/012006

X-bee devices automatically perform the formation of X-bee networks. The coordinator initiates the X-bee network by scanning available channels and PANs in that region. It then picks up an unutilized PAN ID and a channel. Now on a nearby router or an end device can join the coordinator, or the router already joined the same PAN. Each router and coordinator can allow eight end devices to join their device.
All Xbee devices have two different addresses, a 64-bit and a 16-bit address[15]. The 64-bit address is a device address that is unique to each physical device. It is sometimes also called the MAC address or extended address and is assigned during the manufacturing process[8][16]. A device receives a 16-bit address when it joins a Zigbee network. For this reason, the 16-bit address is also called the network address[17]. The 16-bit address of 0x0000 is reserved for the coordinator. All other devices receive a randomly generated address from the router or coordinator device that allows the join[9]. All Zigbee transmissions are sent using the source and destination 16-bit addresses. The routing tables on Zigbee devices also use 16-bit addresses to determine how to route data packets through the network. However, since the 16-bit address is not static, it is not reliable to identify a device[10][13]. So, the 64-bit destination address is often included in data transmissions to guarantee that data is delivered to the correct destination.

5. Experimental setup and results:

Used Xbee modules are initially configured with all the parameters on the XCTU software provided by the DIGI [8]. Mainly, PANID, API mode, and serial transmission parameters are configured to form a mesh network.
In this setup, the communication between the Xbee module and the Atmega board is through the UART. The transmission speed must be set equally in both the controller and Xbee module[9]. For this experiment, both Xbee and controller were configured to a 9600 bauds rate.
Besides many features, the XCTU software has the facility for real-time network monitor and distance tests between the nodes. The range test tool allows evaluating the percentage of lost frame and level of attenuation in the received signal between the communicable nodes. Through the network monitor tab, the real-time deployment of nodes and their interconnections link are monitored graphically. The network, once formed, can be viewed in the network monitor of XCTU, as shown in figure 10 [8].

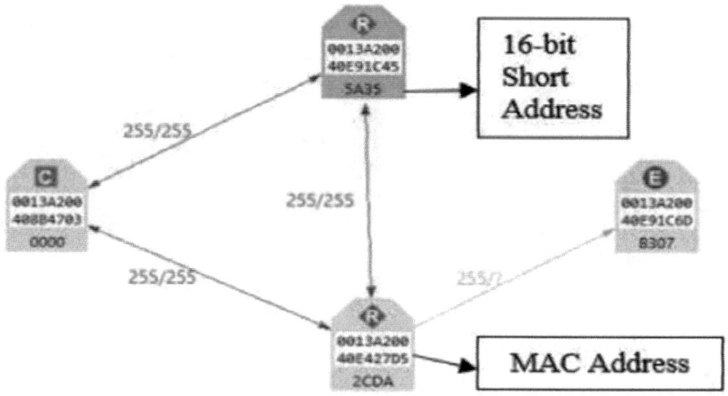

Fig. 10. XCTU network monitor view

5.1. Distance test of functional nodes.

For the realization of the test range between the communicable nodes, two End devices and a coordinator are configured. The coordinator node is mainly connected to a personal computer through the XCTU software. In this case, the test is conducted for two parameters; signal strength test in dBm and verify signal attenuation within the communicable nodes with changes in distance between nodes.

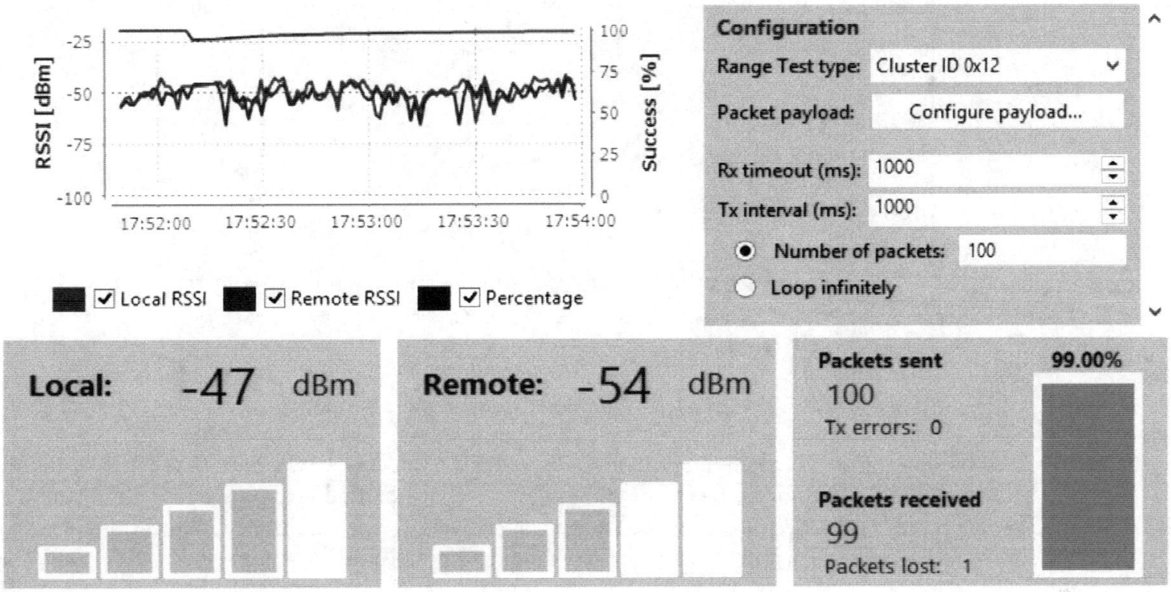

Figure 11. The message transmission test from coordinator to end nodes.

The network is evaluated in two different environmental scenarios, viz. line of sight (LOS) and an obstacle. The LOS condition considers placing the coordinator node and end nodes within the range of 40 meters to 60 meters. And in the second scenario, a concrete wall is counted as an obstacle between the coordinator and end nodes. For each scenario, testing was performed to measure signal strength and to verify transmission reliability. Signal strength is measured using XCTU software, and a frequency of 10Hz latency is considered to evaluate the reliability in transmission data. Figure 11 shows the message transmission test from coordinator to end nodes. The figures include the results of the percentage of success rate and Intensity of signal reached in dBm.

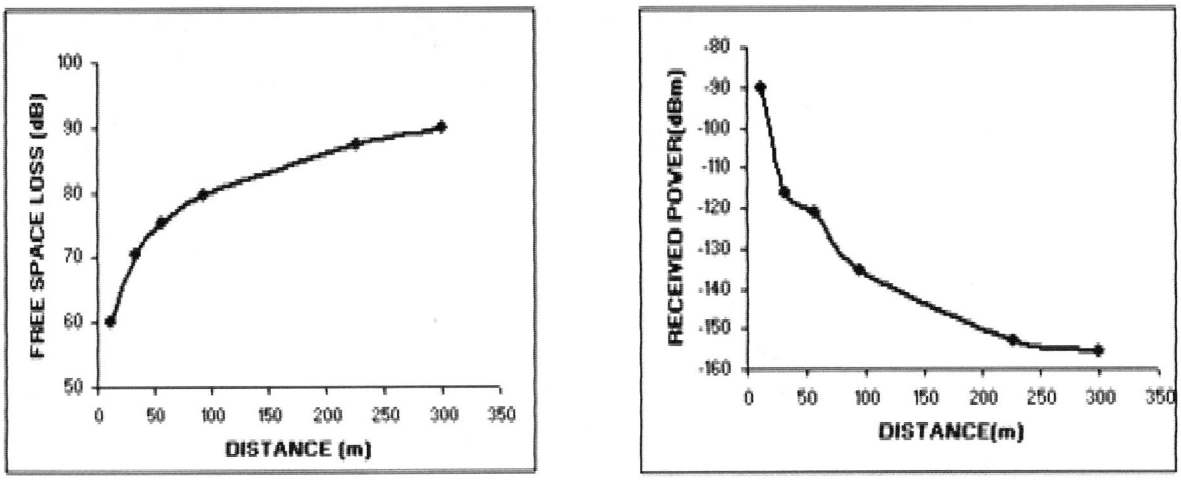

Figure 12. Transmission signal power vs. distance graph

Journal of Physics: Conference Series **2236** (2022) 012006 doi:10.1088/1742-6596/2236/1/012006

Baud rate (bps)	Distance (m) Line of Sight(LOS)	Distance (m) with an *obstacle	Level of the received information
9600	< 40	< 30	Optimal
	41-49	31-39	Stable
	> 50	> 40	Information loss
	**performance changes with an increase in the number of obstacles*		

Table 1. Estimate of the node ranges in the proposed scenario

With the results graph, figure 12, received power vs. distance, we can conclude that with the increase in the distance between the adjacent nodes, the received power decreases or free space loss of the transmission increases as in Table 1. To achieve a wireless sensor network of a stable communication link, the deployment of nodes must be within the line of sight region. It can opt for router nodes if the situation demands. The increase of distance between the neighboring nodes and the obstacles that arise in the distribution of the nodes are indispensable criteria to take into account for the correct design of a Zig-bee Network. Table 1 shows the Estimate of the node ranges in the proposed scenario.

5.2 Field level intensity monitoring:

For field-level testing, three nodes were deployed, connecting four sensors modules to each node, as depicted in fig.5 above. Sensors modules comprise of LDRs and temperature sensor LM35. The resolution of the ADC is 12 bit, and it can be programmed to 8 or 10. The data acquired is continuously fluctuating. By rounding off the values, it could be solved. We have tested these devices without rounding to test the system's accuracy.

Polling period	**NODE A**		**NODE B**		**NODE C**		**NODE Ref.**	
	Lux(A)	**T^{0}C (A)**	**Lux(B)**	**T^{0}C (B)**	**Lux(C)**	**T^{0}C (C)**	**Lux(Ref)**	**T^{0}C (Ref)**
6am-7am	64	26	66	26	66	24	72	28
7am-8am	63	25	67	25	68	25	74	28
8am-9am	69	25	70	25	69	25	82	28
9am-10am	68	25	68	24	68	25	82	28
10am-11am	72	26	71	26	72	24	88	27
11am-12	72	26	72	26	72	26	87	28
12-1pm	73	25	75	25	72	25	81	29
1pm-2pm	74	27	74	27	71	26	88	28
2pm-3pm	71	26	70	26	72	26	88	29
3pm-4pm	71	25	70	25	69	25	88	27
4pm-5pm	72	25	72	25	71	25	87	27
5pm-6pm	64	25	62	25	66	25	79	27
6pm-7pm	37	24	32	24	31	25	56	25
7pm-8pm	6	25	7	25	4	25	6	26

Table 2 . Measured values of Intensity, temperature for shaded and free space area.

Here NODE-A, NODE-B, and NODE-C are deployed in the shaded area of the field and the reference node in the uncovered area. Each node transmits the average of the sensing data. Table 2 shows the data collected by the remote system, and for testing, it represents the average value of the data during the polling period. As per the measured values, the Intensity and temperature are getting less than the data of the reference node. The other type of graphical representation where the control percentage of light Intensity is plotted against polling time is in figure 13.

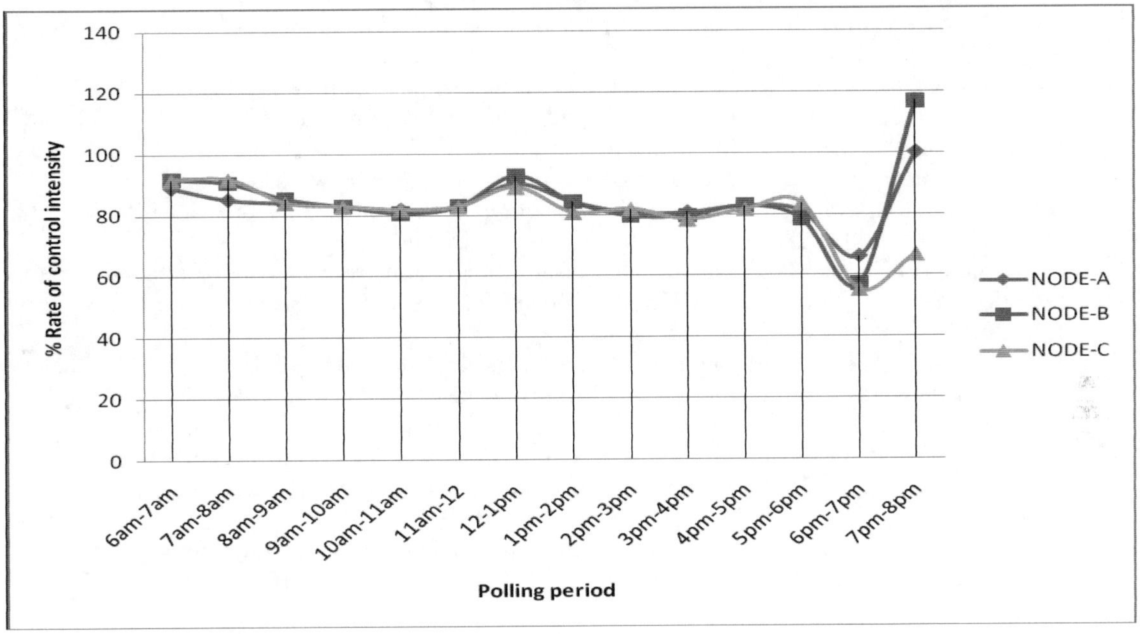

Figure 13. Percentage control Intensity vs. Polling period

6. Conclusion.

While developing such a type of agricultural-based system, the focus is given to implementing a system that works independently and user friendly with minimum human support. The system will be mounted in various areas of the tea field. The reading of the sensors will be stored and manipulated on the remote decision support system as per the requirement of the field and farmer. The number of nodes can be increased easily as the system achieves wireless communication or transmission using long-range wireless Zig-bee. Adding an alarm system or incorporating a GUI (Graphical user interface) system, the purpose system will be more manageable and efficient.

References:

[1] P. B. E. (Tea R. I. of S. Lanka), "Shade Management in Tea Plantations," 2005.

[2] H. Ningombam and O. P. Roy, "A Systematic Approach on Design of Automation System for Tea Farming Based on Embedded Wireless Sensor Network," *Lect. Notes Networks Syst.*, vol. 98, pp. 286–296, 2020, doi: 10.1007/978-3-030-33846-6_33.

[3] M. Redowan, "Management of Shade trees in the tea gardens of Bangladesh with special reference to the Rasidpur and Kamaichora Tea Estate," 2003.

[4] Krishna Kanta Handiqui State Open University, *Scientific Tea Cultivation and Plantation Management*, no. Block 1. .
[5] A. J. Mohotti, G. Pushpakumara, and V. P. Singh, *Shade in Tea Plantations: A New Dimension with an Agroforestry Approach for a Climate-Smart Agricultural Landscape System*, vol. 2. 2020.
[6] "Electrical4u. Light Dependent Resistor." http://www.electrical4u.com/light-dependentresistor-ldr-working-principle-of-ldr/.
[7] "Atmega2560." https://www.microchip.com/wwwproducts/en/ATmega2560.
[8] https://www.digi.com/resources/documentation/digidocs/pdfs/90002002.pdf.
[9] "PS2CTH XBee Pro ZB TH module User Manual Digi." https://fccid.io/MCQ-PS2CTH/User-Manual/User-Manual-2348528.
[10] V. Boonsawat, "XBee wireless sensor networks for temperature monitoring," *Second Conf. ...*, p. 6, 2010, [Online]. Available: http://www2.siit.tu.ac.th/somsak/pub/final_XBeeWSN_100328.pdf.
[11] D. Kominami, M. Sugano, M. Murata, T. Hatauchi, and Y. Fukuyama, "Performance evaluation of intermittent receiver-driven data transmission on wireless sensor networks," in *2009 6th International Symposium on Wireless Communication Systems*, Sep. 2009, pp. 141–145, doi: 10.1109/ISWCS.2009.5285327.
[12] S. Mayalarp, Vachirapol Poombansao, Thanachai Limpaswadpaisarn, Narisorn Kittipiyakil, "Wireless mesh networking with XBee (PDF Download Available)," *2nd ECTI-Conference ...*, no. July, 2010.
[13] A. E. Osorio, E. A. G. Sanchez, G. F. Garcia, S. F. H. MacHuca, and P. S. L. Lozano, "Evaluation of wireless network based on zigbee technology using XBee modules," *2019 IEEE Int. Conf. Eng. Veracruz, ICEV 2019*, 2019, doi: 10.1109/ICEV.2019.8920468.
[14] M. Nesa Sudha, M. L. Valarmathi, and A. S. Babu, "Energy efficient data transmission in automatic irrigation system using wireless sensor networks," *Comput. Electron. Agric.*, vol. 78, no. 2, pp. 215–221, 2011, doi: 10.1016/j.compag.2011.07.009.
[15] J. Y. Jung and J. W. Lee, "ZigBee device access control and reliable data transmission in ZigBee based health monitoring system," *Int. Conf. Adv. Commun. Technol. ICACT*, vol. 1, pp. 795–797, 2008, doi: 10.1109/ICACT.2008.4493875.
[16] "S2DSM XBee S2D SMT User Manual Digi - FCC ID." https://fccid.io/MCQ-S2DSM/User-Manual/User-Manual-2957183.
[17] D. Qi, G. Lu, and X. Dai, "Design of Urban Greening Intelligent Monitoring System Based on Internet of Things Technology," *Proc. - 9th Int. Conf. Intell. Human-Machine Syst. Cybern. IHMSC 2017*, vol. 1, pp. 317–321, 2017, doi: 10.1109/IHMSC.2017.79.

Journal of Physics: Conference Series **2236** (2022) 012007 doi:10.1088/1742-6596/2236/1/012007

Combination of Auto-encoder architecture and super resolution for better segmentation of thinned and cursive handwritten documents

Abstract. In neural networks an auto-encoder architecture has several applications such as image denoising, feature reduction, data compression, image colorization, dimensanality reduction, segmentation and so on. Super-resolution is used to upgrade the low resolution images into high resolution. In order to get a better result on segmentation of thinned hand written images, this paper proposes a method of combination of associative auto-encoder architecture and super resolution for pixel expansion.

Experimental results show that the combination of proposed network and super resolution method accurately segments the thinned handwritten Arabic words.

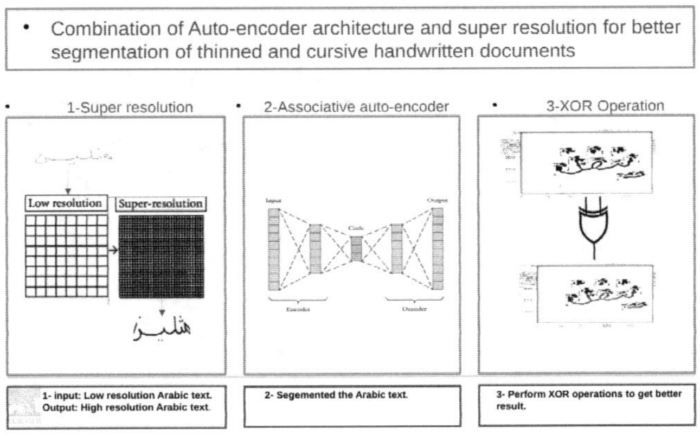

1. Introduction

The segmentation of thinned and cursive handwritten words is most challenging phase, as there is hardly any separation among the adjacent characters as well as characters and words have unusual and varying shapes. This paper deals with Arabic handwritten words as input. Arabic text is cursive in nature and its character representation is somehow difficult. Some of the features of Arabic text are:

* Arabic letters are connected on an imaginary line called the baseline.
* Several Arabic letters are distinguished from each other by the difference in number of dots in different locations.
* The size of Arabic characters varying according to its position in the word.

* Some Arabic letters include a loop that can have different shapes, the shape of the loop changes with respect to the font[1].

The main aim of this paper is by using associative auto-encoder architecture on Arabic text images(Plain text and corresponding ground truth images), creating a segmentation system for Arabic handwritten text, then a thinned hand written Arabic text is given as input for testing. This thinned input image is modified using pixel expansion of super resolution methods..

2. Research highlights

* Conducting the pixel expansion by using super resolution of thinned letters.
* Associative auto encoder architecture is used for segmentation.
* Bit-wise XOR is performed on the predicted images.

3. Methods

3.1. Auto-encoder

Autoencoder is an artificial neural network that use encoder and decoder to represent raw data in a compressed form. It is an unsupervised method[2].

An auto-encoder has three parts named encoder, code and decoder as shown in figure 1[3].

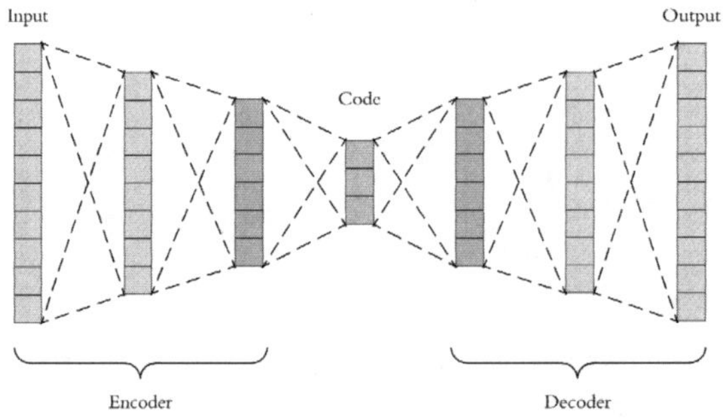

Figure 1. Architecture of associative auto-encoder.

In auto-encoder the input data pass through the encoder to produce the code. The decoder receives the data as input produced by the code layer. Here the layers of input (encoder) and out put (decoder) need to be the same in numbers.

3.2. Super Resolution(SR)

Super resolution (SR) is used to upgrade the low resolution images into high resolution images.

In the proposed work, the pixels of thinned Arabic input text image are expanded using these SR methods and then it is given to the associative auto-encoder architecture for testing.

4. Algorithm

The method used in this work is as follows:-

(i) 90 Gray scaled images of Arabic text and its corresponding ground truth images are scanned(figure2)

(ii) Resize and reshape original and its corresponding ground truth images.

(iii) Create input image named input_img with size (248,248,3).

(iv) Encoder part starts with creating a convolution layer named conv1 with maxpool and end with a convolution layer named conv3.

Journal of Physics: Conference Series **2236** (2022) 012007 doi:10.1088/1742-6596/2236/1/012007

(v) Corresponding decoder part creats.
(vi) Create a model named auto-encoder using mean square error and for optimization using RMS (Root Mean Square).
(vii) With batch size 4 and epochs 150 trained the system.
(viii) Super resolution applied on the thinned input image.
(ix) Result of step(viii) used for prediction.
(x) Bit-wise XOR performed on the predicted image for segmentaion.

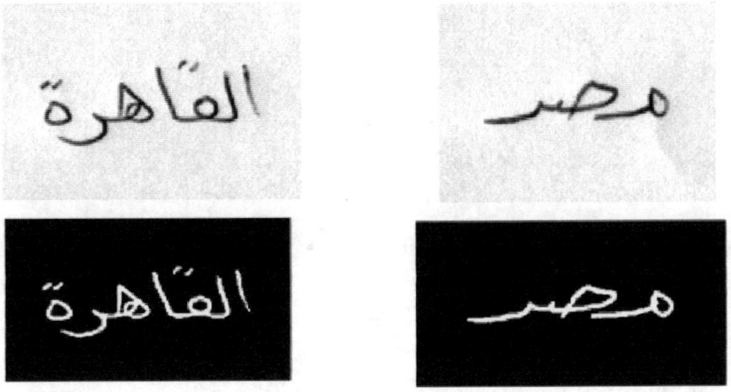

Figure 2. Sample example of data set and corresponding ground truth images.

5. Experimental Result

We run the code on 90 gray scale images of Arabic text and its corresponding binary images (ground truth)[4].

A thinned Arabic handwritten text image (figure 3) is given as input after pixel expansion using super resolution method for prediction. We get a better result as shown in figure 4. For segmentation we perform bit-wise XOR on the predicted image. The result shown in figure 5.

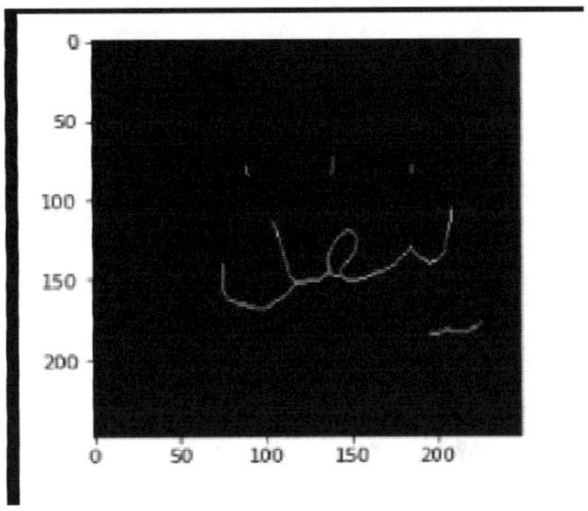

Figure 3. Thinned input image.

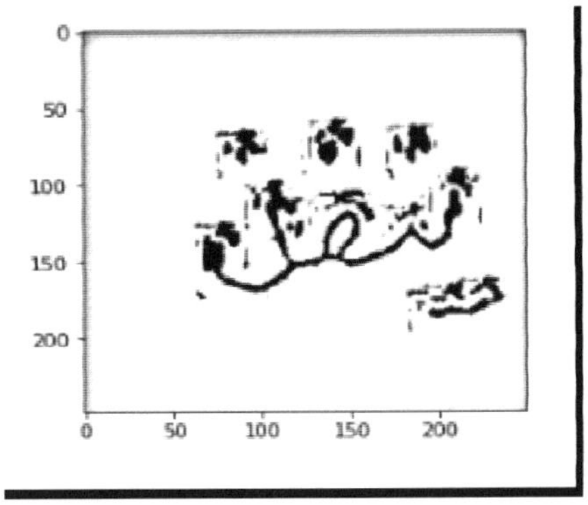

Figure 4. Resulted image after prediction.

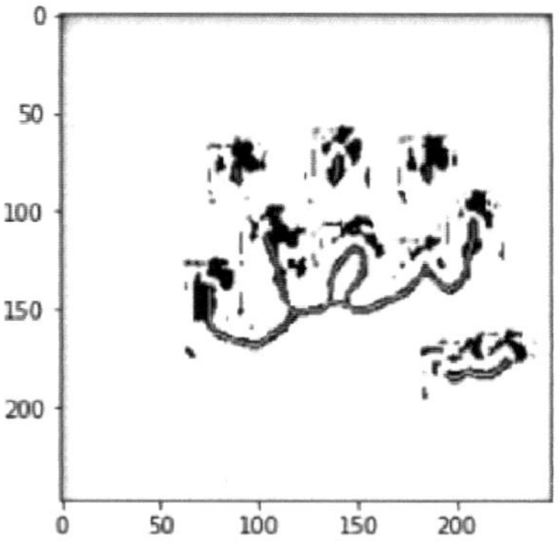

Figure 5. Segmented image after bit-wise XOR.

6. Conclusions

This paper presents a combination of associative auto encoder and super resolution methods for segmenting the thinned Arabic handwritten text. We used the [3] data set of 90 images for training purpose. A thinned handwritten image from outside of data set is used for testing purposes. It gives a better result on these image.

In future work we can escalate the performance of the proposed system by using more data sets on the same script as well as on other cursive scripts.

References

[1] A.E. Hassanien et al. (Eds.): AMLTA 2014, CCIS 488, pp. 11–17, 2014. © Springer International Publishing Switzerland 2014.
[2] https://www.mygreatlearning.com/blog/autoencoder.
[3] https://miro.medium.com/max/ 700/1*44eDEuZBEsmG_TCAKRI3Kw@2x.png
[4] https://www.bl.uk/about-us/terms-and-conditions

Journal of Physics: Conference Series 2236 (2022) 012008 doi:10.1088/1742-6596/2236/1/012008

Internet of Things for sustainable urbanism

Paribhasha Sharma[1] and Srinivas Jangirala[2]
[1]Associate Professor, Jindal Global Business School, O.P. Jindal Global, Sonipat, Haryana, India
[2]Associate Professor, Jindal Global Business School, O.P. Jindal Global, Sonipat, Haryana, India

E-mail: [1]paribhasha@jgu.edu.in, [2]sjangirala@jgu.edu.in

Abstract

Cities around the world are facing enormous strain to sustain and improve the quality of life (QoL) owing to rapid urbanization and rising populations. Management of urban resources in a responsible manner is key to sustainable development in rapidly urbanizing regions. Cities are increasingly making use of modern technologies with a focus on cost reduction, optimal resource utilization and creation of more liveable urban environment. Such cities, called smart cities, have gained traction with policy makers, politicians and urban managers having the attributes of sustainable urbanism, QoL, and smartness. Smart cities provide digital intelligence to existing cities by creating a ubiquitous, integrated and smart environment where IoT applications impart seamless interconnection, interaction, control and insights about the isolated systems within the cities. This paper discusses and reviews the role of IoT for sustainable smart cities by highlighting IoT applications for smart cities. The challenges and opportunities associated with IoT enabled smart cities are also highlighted.

Keyword: Sustainable urbanism, smart cities, IoT, privacy and security, QoL

1. Introduction

Majority of the world population now inhabits urban spaces, with 55 percent of the people living in cites in 2018 and an estimated 70 percent of the world population is set to become urban by 2050 [1]. The most urbanized regions of the world are still in the West with more than 60 percent of their population living in cities. However, other regions of the world particularly Asia is also catching up with the West in this urbanization trend with approximately 50% of its population living in urban spaces. Many of the global cities are experiencing enormous pressure to sustain and improve the quality of life due to rapid urbanization and rising populations. Cities contribute a greater percentage to the Gross Domestic Product (GDP), provide employment, and increase per capita spending, thus driving the country's economy. Cities offer better services, but they also harbour enormous problems like degraded quality of the environment, deficient basic infrastructure and services, increasing exclusion and feminization of poverty. Management of urban resources in a responsible manner is key to sustainable development in rapidly urbanizing regions.

The phrase "smart city" is a new concept that has gained popularity lately. The word was coined by governments, institutions, and citizens, as well as academic and industrial sectors [2]. This new paradigm, which is becoming increasingly popular and appealing, has a lot of definitions. A smart city, for example, may be defined as "a well-defined geographical area in which high technologies, such as Information Communication Technologies (ICT), logistics, energy production, etc. work collaboratively to create benefits for citizens in terms of well-being, inclusion and participation, environmental quality, and intelligent development; it is governed by a well-defined pool of subjects, able to state the rules and policy for the city government and development" [2].

More recently, in combination with an early effort at actual implementation, the smart city idea had been criticized of being overly technology-centred and mostly pushed by the technology enterprises and without any genuine attention to the municipalities and the requirements of people. Consequently, a more sustainable strategy is overwhelmingly necessary.

Internet of Things (IoT) refers to interconnection of things, people as well as objects, to the internet and to other connected devices that enables data collection and sharing. IoT interconnectedness can increase society's ability to build smart and efficient cities. It can analyse and solve problems in architecture, agriculture, safety, surveillance, sanitation, etc., and solve the sustainability concerns. It offers a balance between sustainability issues through processing and information sharing. It also helps communicate the demands and problems of every area with the help of technology sensors. By allowing things to cooperate, strengthening decision-making through improved data collection and improving the surroundings, IoT technology can enhance our effectiveness by reducing resources, applying them and distributing them better. This paper discusses how IoT solutions can be applied to sustainable smart cities. The rest of the paper is organized as follows: Section 2 provides an overview of sustainability, sustainable development for cities and IoT. Section 3 examines the functional capacity of IoT to create smart cities. Section 4 discusses the challenges and opportunities of urban IoT with emphasis on security. Section 5 presents IoT use cases in selected cities of the world. Section 6 concludes the discussion.

2. Sustainable development, Smart Cities and IoT

Sustainability has a long history and is widely accepted, contrary to the idea of smart cities (it was initially presented in 1987). It is based on three major elements or dimensions: social, ecological and economic. A "sustainable city" is an entity which, according to a more modern definition, does not transcend the capacity of the surrounding city's resources and energy resources as well as trash disposal. In practice, the urban resource consumption in a city should not exceed or correspond to the amount given by the natural environment to achieve its objective (e.g., energy, soil or water resources). In addition, the environmental capacities of resorting to residents and other ecosystem members should not surpass the pollution level produced by municipal operations. The notion of sustainability has been challenged to be outdated and inappropriate to the demands of the highly digital world of today, which despite its simplicity and rationalization is characterized by extremely quick changes and growth.

Recognizing cities as the transformative force for sustainable development, the United Nations established a stand-alone and dedicated sustainable development goal on human settlements (SDG 11) that sets out to "make cities and human settlements inclusive, resilient and sustainable" [3]. This goal is reflective of the need for sustainable measures for regeneration of cities in sustainable forms. SDG 11 devotes a significant amount of attention to people and to the common good [4]. There are seven outcome-oriented (refer to figure 1) and three process-oriented targets in SDG 11. The goal of SDG 11 is "to provide safe and affordable housing and public transport and develop well-planned cities with environmentally sustainable buildings and increased public spaces were cultural and national heritage is protected. It also aims to improve resilience to disaster and risk management" [5]. SDG 11 provides an opportunity for rethinking urban planning along the three dimensions of sustainability – social, environmental and economic [6]. Sietchiping et al. [7] argue that SDG 11 can help bring sustainability to the fore of urban planning and development agendas.

Figure 1. Targets of Sustainable Development Goal (SDG) 11.

It is also observed that billions of gadgets and systems are integrated into the infrastructure of a city. These include end-user equipment, municipal lighting systems, road traffic and pedestrian management, monitoring of water and gases, structural monitoring and waste management, and smart medical services, to name a few. In order to improve the safety, efficiency, production and quality of life of its citizens, the city is therefore de facto required to employ ideal computers and communication, integrate, manage and analyse the massive data. Smart city applications aid inhabitants and a basic environment that includes smart economy, smart government, smart individuals, smart mobility, smart environment and smart life [8].

"Though, a formal definition of a smart city exists" [9], from a data management viewpoint, Gharaibeh et al [10] suggested that "a smart city employs a combination of data collection, processing, and disseminating technologies in conjunction with networking and computing technologies and data security and privacy measures encouraging application innovation to promote the overall quality of life for its citizens and covering dimensions that include: utilities, health, transportation, entertainment, and government services".

The evolution of these concepts led to academics suggesting a new paradigm, dubbed "smart sustainable city,". More specifically, this paradigm tries to combine a municipality's urban sustainability with smartness, emphasising the importance of considering both factors together in order to achieve the best urban governance. According to a comprehensive definition [11], smart sustainable cities are defined as "innovative cities that use ICT and other means to improve quality of life, the efficiency of urban operations and services, and competitiveness, while ensuring that they meet the needs of present and future generations in terms of economic, social, environmental, and cultural aspects".

It is crucial to assess the effects of the usage of such intelligent devices on individuals' everyday life. This is based on new opportunities offered by today's wave of technological innovation particularly on the rising application of Internet of Things (IoT) devices and entities. The IoT is a modular approach to interconnect various devices over the internet to enable them to accumulate and exchange data with each other [12]. IoT has been variously defined [13-16] however, Gubbi et al [17] provide a more user centric definition. According to them, "Internet of Things for smart environments is interconnection of sensing and actuating devices providing the ability to share information across platforms through a unified framework, developing a common operating picture for enabling innovative applications. This is achieved by seamless large-scale sensing, data analytics and information representation using cutting edge ubiquitous sensing and cloud computing" [17].

Smart sustainable cities may enhance many elements of their urban administration through the launch of IoT innovations (as illustrated in figure 2), such as city transport, environmental monitoring, public transit, e-government, public lighting, safety and security. The IoT technology in sustainable smart cities should enable monitoring, control and management of all accessible resources such as water, soil, electricity, people and others.

Figure 2. Essential issues in a smart city. (Source: Adapted from Belli et al. [2], 2020)

The next section examines how IoT functional capacity can help in the creation of smart cities.

3. Functional capacities of IoT to generate smart cities

IoT technologies can produce smart cities in six areas: "smart economy, smart people, smart government, smart mobility, a smart environment and smart living" [18]. Smart cities can connect objects, people and information through computer networks via IoT technologies. In order to deploy IoT technology correctly, sensor-related issues, such as accuracy, bandwidth and cloud computing need to be maintained constantly. Data receivers may influence samples of data collection, numerical variables, and conclusions on infrastructure. The Internet of Things technology comprises of thousands of networked nodes which like a system of functionality, cooperate concurrently, transforming the environmental adaptations into compressed forms of data. IoT technology classifies data via ubiquitous cloud computing through Internet connectivity extension through communication with devices and common items. Many scientists have tried to describe the nature of the IoT. Through different data compilation processes, objects can operate like small computers. Logical patenting, obvious with machine learning, can be predicted and information learned from sensors can be signalled over wireless cloud platforms. IoT technology is important because it can promote smart cities' development and wealth.

As described in section 2, IoT constitutes one of the fundamental technologies for the development of a smart city that involves a steadily growing number of linked devices in various cities. The overview of IoT Analytics' projected number of linked IoT devices in 2015-2025 (also known as smart objects). As is evident from figure 3, the number of IoT devices have been increasing since 2015. This increasing trend is projected to continue till 2025, therefore, reflecting that communication infrastructure needs immediate and urgent attention in smart cities as connectivity anytime and anywhere is what defines a smart city.

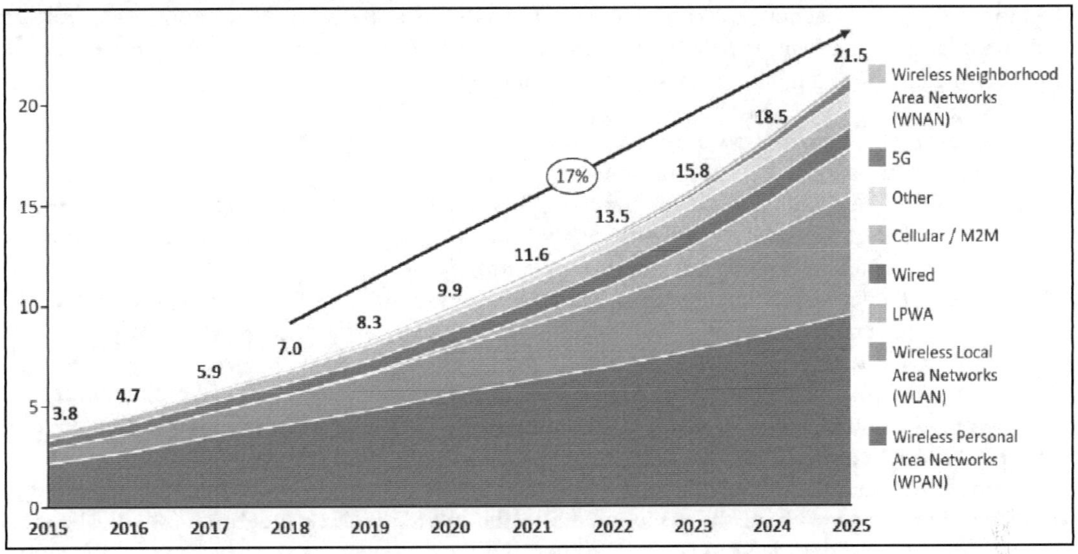

Figure 3. Trend on the global number of connected IoT devices in the period 2015-2025 [19].

Note:- No PCs, laptops, landline telephones, cells or tablets are included in the IoT connectivity. There are active nodes / devices or gateways, not every sensor/actuator, concentrating on end-sensors. Simple one-directional communications technology not considered (e.g., RFID, NFC). Wired includes Ethernet and Fieldbuses (e.g., connected industrial PLCs or I/O modules); Cellular includes 2G, 3G, 4G; LPWAN includes unlicensed and licensed low-power networks; WPAN includes Bluetooth, Zigbee, Z-wave or similar; WLAN includes Wi-fi and related protocols; WNAN includes non-short-range mesh; Other includes satellite and unclassified proprietary networks with any range.

4. Smart Cities: Challenges and opportunities

In this section, we discuss the challenges and opportunities in smart cities.

4.1. Challenges. Although smart cities offer a wide range of benefits to its citizens, there are many hurdles to digital transformation in metropolitan areas. IoT related security is one of the most significant challenges which has been elaborated upon in the latter part of this section. Here's a summary of some of the most urgent difficulties in the development of smart cities:

4.1.1. Infrastructure: Although our everyday life has been greatly affected by the emergence of new technology, the infrastructure in most cities has remained untouched since the nineteenth century in general. Smart city projects need a strong foundation to thrive on, understandably. In particular, IoT sensors capturing different data from air pollution to traffic congestion levels require modern infrastructure supported by novel technology in order to completely deliver their worth. In most developed cities, various infrastructural difficulties such as water and steam pipes, Internet broadband, energy etc. still exist.

Thus, the successful infrastructural modifications depend on the allocation of intelligent resources, generous finance and complete government assistance.

4.1.2. Privacy Concerns. Most people want to profit from smart city projects and benefit from greater safety, lower crime rates and better living standards. The price to pay, though, many feel is invasive and maybe menacing. Monitoring cameras can lower the rate of speeding vehicles and other offences in the city, but these cameras also create the feeling "big brother watches", thus, generating fear. The volume of personal data collected by IoT smart sensors is a matter of grave concern for citizens. Full transparency of the usage of data as well as education efforts to inform citizens of the way in which smart cities function, ought to at least partly assuage these worries.

4.1.3. Smart cities security issues. In smart cities, security concerns arise because many individuals are suspicious about smart city programmes. IoT devices are basically safety gaps. The increasing quantity of IoT sensors and the expanding interdependence of interdependent urban silos poses legitimate issues. If the safety standards remain unchanged, cyber criminals might shut down a whole city. Fortunately, technology companies are inventing security solutions based on large-scale analysis, blockchain and encryption technology that are meant to manage ever more advanced cyber threats. In order to eliminate risks, smart city developers invest in these latest generation security technologies.

4.1.4. Social inclusion. Examples of adequately implemented efforts could be beneficial in assuring inclusiveness while developing smart cities. If inclusiveness is not achieved, the best intentions could possibly be negated. For example, because many people do not know how to utilise the technology, a town might not undertake a health project for the elderly people. In order to promote social inclusion and to communicate with all the stakeholders in the city, smart city initiatives should be carried out. Besides the above-mentioned challenges, other challenges related to citizens, mobility, environment, and governance was forwarded by Belli et al [2]. These can be specifically sub-categorized into various ground level challenges. The details of this classification are given in table.1.

Table 1. Classification of Smart city challenges.

Citizens	Mobility	Environment	Governance
• Unemployment • Social cohesion • Poverty and inequality • Aging population • Data safety and privacy • Urban violence and insecurity • Shortage in access to technology • Health and emergency management • Deficit of social services	• Sustainable mobility • Inclusive mobility • Multimodal transport system • Interoperability • Traffic congestion • Alternative mobility • Lack of public transport • Weak ICT infrastructure • Pollution	• Energy saving • Holistic approach to environmental and energy issues • Climate change effects • Rapid population growth • Energy resources scarcity • Water scarcity • Pollution	• Flexible governance • Territorial cohesion • Interoperability • Combination of formal and informal government • Gap between citizens and government • Unbalanced urban development • Deficit of social services • Unemployment • Sustainable economy • Weak ICT infrastructure • Shortage in access to technology

Source: Adapted from Belli et al.[2] (2020)

4.2. Opportunities

Though a number of challenges exist for smart cities, they also offer numerous opportunities, some of which are discussed below:

4.2.1. IoT management. IoT needs a safe, efficient architecture to improve the collection of urban data. Omnipresent and cooperative, smart urban sensing can offer a smart environment. Others are unavoidably not controlled, such as latencies and packet loss. One idea is to "employ the MANET nodes as mobile relays to swiftly gather urgent data from wireless sensor networks without sacrificing battery life" according to the coordination protocol of Mobile Ad hoc Networks (MANET)[20]. Their cluster generation process, according to simulation results, is trustworthy and reliably transmits more than 98 percent of packets in street and square circumstances [20]. Other issues, such as the intersection of IoT and intelligent transportation systems, require greater investigation.

4.2.2. Data management. In a smart city, data is crucial. Smart cities will generate vast amounts of data that will be difficult to understand, manage, and analyse. On the other hand, mobile phone data can help achieve a variety of smart city objectives. Data from mobile phones can be used to create a variety of urban applications. Mobile data, for example, might be used to predict the volume of traffic and other mobility demands. It is possible to perform a transportation analysis. Data from mobile phone sources and data from taxis' Global Positioning Systems can help manage transportation resources in the future, such as public taxi demand. A smart city requires a lot of data storage, which can be done with a metasurface or a bunch of nanoantennes. This technology has the potential to have an impact on a variety of fields (such as encryption, imaging, data storage, and communication). The response of a given smart city to unknown facts is critical, as is the topic of how to gauge a smart city's strength.

4.2.3. Smart city assessment framework. The city's quality of life, including health, safety, and wellness, must be monitored. Using an evaluation framework, several characteristics such as smart city initiatives and stakeholders' interests must be taken into account [20]. Its goal is to compare and analyse the characteristics of various smart cities in order to identify new issues, measure advantages, and assess performance.

4.2.4. Vehicular ad hoc network (VANET) security. In smart cities, a major need for VANETs is efficient security support. One aspect is how they may be secured by developing solutions which minimise the possibility of network assaults and can also lessen the impact of a successful attack. In the field of authentication and driver analysis, certain security issues exist. A smart city requires "lightweight, scalable authentication frames to protect drivers from external or internal attackers" [20].

4.3. IoT Smart cities security

Although the cities seek to be smart through the adoption of various solutions, smart city applications emphasize a few privacy and safety concerns, as city operators are expected to defend the information generated or gathered from unauthorized access [21].

Information, communication and physical worlds should satisfy core security and data security standards, including confidentiality, integrity, non-repudiation and accessibility, access control and privacy [22]. In addition to these fundamental needs, the protection of an intelligent town nevertheless faces a series of specific problems. On the one hand, an intelligent city collects granular and confidential information that is sensitive to people's lives and environment; on the other it analyses and manipulates this information and influences people's lives. Some of the major security issues related to IoT in smart cities include:

4.3.1. Privacy Leakage in Data Sensing. A smart city is subject to private leaks and information provided by external adversaries since data is gathered, transferred and processed in private cities. In a smart city, disclosed privacy can contain the identity and location of the user in the transport industry, health status in healthcare, intelligent surveillance-driven life style, smart power, house and community etc. It would be huge supervision to divulge this information that is important to privacy in both the physical and communication realms to unauthorized entities.

In order to protect user privacy during data collection, several security and privacy measures, including encryption, anonymity and access reserve can be applied. Martínez-Ballesté et. al. [22] propose to use smart city applications with a range of privacy principles and criteria. They propose a five-dimensional model of citizens' privacy in smart cities including identity privacy, query privacy, location privacy, footprint privacy and owner privacy to tackle generic difficulties related to data breach. However, certain private information can still be exposed inadvertently to unreliable entities.

Smart surveillance, for example, may record indices, styles or even privacy of local inhabitants, while originally meant to monitor crime in actual and virtual circles. In an intelligent home, a camera also detects burglary or odd occurrences using a monitoring camera. Private information about your home area may be obtained by infiltrating attackers in a smart house which harms the privacy of the residence. Majority of current security and privacy protection against external assailants and attackers have been addressed. However, prospective attackers inside the community, including agencies, staff, and security guards, who may access monitoring data can either steal information from consumers or leave a loophole for external attackers. In addition, the data in a smart city are of high granular size and of various type that vary with privacy needs. The development of customizable methods for protecting privacy in a smart city is a challenge to reconcile the compromise between privacy and efficiency [23].

4.3.2. Privacy and Availability in Data Storage and Processing. Because a smart city has the capacity of cloud servers to store and analyse data across information, cloud servers confront security threats. If smart city data are in plain texts, they are disclosed immediately to the cloud server during storage and processing. An option would be to encrypt the intelligent city data and transfer ciphertexts for storage and processing to the cloud server. Whilst the untrustworthy cloud server prohibits direct access to information collections, it is not possible for the cloud server to handle encrypted data and carry out effective analysis for intelligent urbs. The newest advance in completely homomorphic encryption illuminates processing by encrypted data, for instance, summing and comparing. Another difficulty in terms of performance is the processing overhead, especially when mass data is used in a smart city.

The data sharing and access management is another difficult problem to secure a smart city. For instance, data from the road can be gathered in the form of a crowdsourcing via placed cameras or cell phones and GPS. It is difficult to set access policy during global road planning and to facilitate the private sharing of data across employees. Smart city data storage and exchange therefore require a lot of research.

4.3.3. Trustworthy and Dependable Control. The intelligent city with double-way control depends on the control system and on the drive systems for conducting the operations of the control centre. The systems to regulate and feedback, especially in public and industrial infrastructure, in the physical world are becoming more appealing targets for terrorists, attackers, and criminals. Spoofing attacks, denial of service assaults, and malicious injections of data would disrupt the city of smart business, which would be impaired in management, control and operation [24].

Most malicious attempts and misbehaviours are identified via inspection and auditing by third parties. In Akhunzada's [25] work, the capabilities of data integrity and digital signatures are used for data integrity, access control, and more in software-defined networks. Confidential computing is a state-of-the-art approach to resist modifications of operating system and applications. But these systems need high delay and high false rates in order to identify "intelligent" assaults in an intelligent city. As control reliability is regarded the highest priority in a smart city, it becomes harder to identify hostile assaults and misbehaviours, necessitating a coordinated effort across many parties involved.

The next section examines the application of IoT in smart cities.

5. IoT use cases for smart cities

Journal of Physics: Conference Series **2236** (2022) 012008 doi:10.1088/1742-6596/2236/1/012008

Smart cities can be made sustainable by making a better use of public resources, offering better quality services to the citizens while reducing the operational costs of public administration [26]. IoT applications can help achieve this goal by creating cost-effective municipal services, enhancing public transformations, reducing traffic congestion, keeping citizens safe and healthier [27]. As IoT uses the internet for merging different things, it is important that all existing things get linked to the internet for remote usage and monitoring. Some of the major applications of IoT for a smart city is illustrated in figure 4.

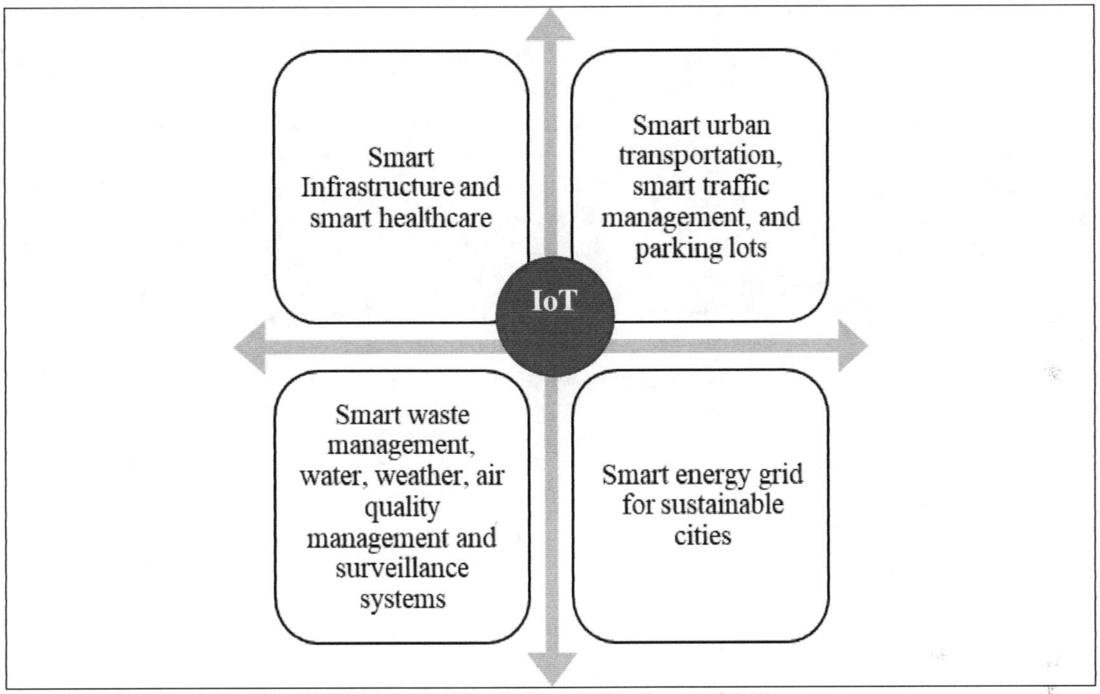

Figure 4: The main applications of IoT

5.1. Major applications of IoT for a smart city

5.1.1. Smart Infrastructure and smart healthcare. Smart infrastructure includes smart homes, offices, schools, data centres, factories, ware houses, etc. Application of IoT technologies for smart buildings can help manage security, surveillance, automated operations, energy management, and so forth distantly. Smart homes, smart offices, smart warehouses, and other smart buildings perform their tasks efficiently and accurately [28]. For example, a smart home uplifts the quality of life of its inhabitants by efficiently and effectively controlling home appliances, energy consumption, surveillance, lighting and temperature control, etc. [29-32]. Sensors used in smart appliances and systems help in checking the environment state and then taking suitable action like dimming lights or switching off air conditioner. Such smart systems also help in demand prediction. Similarly, the productivity of supply chain management can be enhanced using smart warehouses [33]. The most significant benefit of smart homes and buildings is convenience to the users as they are freed up top perform other responsibilities.

Sustainable smart cities should have smart healthcare systems in place. IoT applications can be used in medical care by improving medical information systems. "Innovative data capture offers continuous and ubiquitous medical device access from any connected device over the internet" in IoT-based healthcare [34]. Tracking people and things, such as patients, medical personnel, and ambulances, identification of individuals, and automatic data collecting and sensing are just a few of

Journal of Physics: Conference Series **2236** (2022) 012008 doi:10.1088/1742-6596/2236/1/012008

the benefits of IoT-based healthcare in smart cities [27]. Citizens can avail online appointment, online doctor's consultation, monitor patient vitals using wearable sensors, and other medical services. Patient tracking in clinics and hospitals helps in providing faster and better services, object tracking (ambulance, blood, medicines, etc.) help in checking their availability. Wearable sensor devices by patients provide real-time information on their health indicators and thus, help in getting timely medical advice and support. Various mobile health (m-health) platforms offer health services for remote monitoring of patients and communication between professionals, relatives and patients. m-health platforms are user-oriented, can be personalized and allow easy access to a number of services and knowledge [35].

5.1.2. Smart urban transportation, smart traffic management, and parking lots. Urban transportation system can benefit significantly by IoT applications. Smart urban transportation system includes among others automatic number plate recognition, vehicle counts on the roads, traffic signal automation, smart lighting and smart parking. Application of IoT in managing vehicular traffic information can help in real-time traffic management, thereby, benefitting the citizens, urban governments as well as the urban environment. A combination of sensing capabilities, GPS installed modern vehicles, air quality and acoustic sensors along a given road is of great help in traffic monitoring and in making the cities sustainable [36].

Lack of proper parking in cities is a major concern for the citizens. Smart parking lot applications can monitor the number of vehicles as well as their arrival and departures in different parking lots of the city. Usage of road sensors and intelligent displays help the drivers find the best path for parking in the city. Smart parking lots offer a range of benefits to the users, vendors/contractors of the parking lots, the government and the general public. Finding parking lot faster means less vehicular pollution, lesser traffic congestion, and happier citizens [27, 37].

5.1.3. Smart energy grid for sustainable cities. Smart grid refers to the integration of traditional electrical power grid with latest telecommunication and information. IoT is one of the most recent enablers for smart grid technologies [38]. In a variety of circumstances, the Internet of Things may be used for cost-effective power generation, management, transmission, and consumption. Intelligent and autonomous controllers, advanced software for data management, and two-way communications between power utilities and customers are all part of an automated and advanced energy distribution network created by an IoT application for smart grid [27].

5.1.4. Smart waste management, water, weather, air quality management and surveillance systems. Quality of life in a smart city is determined not only by smart physical entities like infrastructure and other facilities but also by its environment. A clean and healthy environment is the one of the pillars of a sustainable smart city. Most of the cities around the world face the challenge of managing their waste which is generated daily and mostly end up in landfills at the outskirts. Urban waste management includes processes such as collection of waste, disposal, recycle and recovery, managing and monitoring of water materials [39]. Use of IoT in waste management may result not only in significant savings but ecological advantages as well. IoT can "connect the end devices", such as intelligent garbage containers, to a control centre, where data is processed and the best way to run the collection truck fleet is determined using optimization software [26].

Similarly, water and weather systems can utilize diverse sensors in an IoT-based environment for data management related to water distribution, temperature, rain, solar irradiation, and wind speed. The data can be analysed to forecast rains and other weather conditions including floods, detect leakages in water distribution systems, enable the urban governments to develop innovative methods to plan and manage scare resources like water through IoT. Air quality can also be monitored using urban IoT. This can be done by deploying pollution and air quality sensors across the city and making the data available to the citizens on a real-time basis.

One of the most critical factors in any smart city ecosystem is the issue of security and privacy. It must be ensured that every entity in a smart city is secured at all times and therefore, constant monitoring and observation is needed. Conventional security techniques like CCTV systems do not have the capability for intelligent data processing. IoT applications can create smart surveillance systems that can raise alarm in case of any security lapse and help in crime reduction, improvement in city management, crime detection and resolution.

5.2. Urban IoT applications around the world

IoT applications for sustainable urban management is gaining traction with policy makers, businesses as well as individuals. Widespread diffusion of IoT backed by popular demand and technology advances could lead to an invaluable contribution to the economy. Some examples of urban IoT applications in the world are given below:

5.2.1. London (United Kingdom). London is well recognised for its excellent passenger management and transit networks. Congestion management based on number plate identification was effective in minimising traffic bottlenecks during peak hours. Wi-Fi connectivity on the Tube, automated road management, and cycle-rental programmes are among the numerous IoT-based smart applications[28]. London applies IoT for smart waste management under the London Plan. The data is connected to a centralized server for real time monitoring and management.

5.2.2. Santander (Spain). The largest testbed for IoT deployment is the SmartSantander project in Santander city in Spain where 12,500 sensors have been deployed across the city. These sensors use an IoT network to monitor the environment, the number of pedestrians, available parking spots, and the remaining volume of garbage bins, among other things, in order to establish smart transportation and smart communities [28].

5.2.3. Nice (France). The city of Nice in France is using the IoT to digitize existing services and create new ones. The French city has designed an eco-city zone for a sustainable smart city, collecting varied data and sharing the processed data with the citizens with an aim to improve quality of life in Nice[2]. Nice employs IoT-services for urban mobility, energy and environment monitoring and waste management.

5.2.4. San Francisco (United States). San Francisco utilizes IoT technology to improve operational performance of buildings, extend transportation systems, centralize waste management procedures, and reduce energy consumption. Thus, San Francisco has smart energy, smart community, smart transportation and smart community serving its citizens.

5.2.5. Barcelona (Spain). Barcelona has achieved a wide range of benefits of a smart city by utilizing IoT for urban systems such as LED-based street lighting, waste disposal with the use of smart bins, smart public transportation including cycles and bus transit systems, low-cost and easy-to-use sensors to detect noise pollution, open access data, and smart water management systems.

6. Conclusion and significance

In this paper, we have investigated the sustainable smart cities and Internet of Things (IoT) and discussed how IoT technologies interconnectedness can increase the society's ability to build intelligent and efficient cities. Next, we defined the goals for urban development and sustainable development such that the smart cities planning can be achieved with proper regulations and infrastructure. We further focused on the smart city challenges and opportunities, where the challenges in the development of smart city have been categorized. We paid particular attention to the issue of safety and privacy in upcoming smart city applications. We have examined and emphasized factors such as data sensing privacy leakage, privacy and data storage and processing accessibility, reliable control and reliable control. In addition, we presented case applications for smart cities focusing on smart infrastructure, intelligent parking, intelligent trash management and air quality control.

This paper is particularly significant for practitioners and public policy experts including government. It sheds light on how IoT can be employed for an inclusive sustainable city. The concerned government authorities can use the insights of this study in effective implementation of smart cities to make our cities sustainable. We expect this paper will provide further insight into the safety and privacy of intelligent cities, which will show a further pioneering research effort along this new path.

References

[1] United Nations Department of Economic and Social Affairs. *World Urbanization Prospects. The 2018 Revision*. New York, NY, USA: UN Population Division. (2018). Available from https://population.un.org/wup/Publications/Files/WUP2018-Report.pdf.

[2] Belli L., Cilfone A, Davoli L, Ferrari G, Adorni P, Di Nocera F, et al. IoT-Enabled Smart Sustainable Cities. Challenges and Approaches. *Smart Cities*. 2020; 3(3): 1039-71. Available from https://doi.org/10.3390/smartcities3030052.

[3] United Nations General Assembly. *Resolution 70/1. Transforming our world: The 2030 Agenda for Sustainable Development—A/RES/70/1*. United Nations. (2015). Available from http://www.un.org/ga/search/view_doc.asp?symbol=A/RES/70/1&Lang=E.

[4] Satterthwaite D, Bartlett S. Urbanization, development and the sustainable development goals. In Bartlett S, Satterthwaite D, editors. *Cities on a finite planet: Towards transformative responses to climate change*. 1st ed. Oxon and New York: Earthscan from Routledge. 2016. p. 1–16.

[5] United Nations. *A short guide to Human Settlements Indicators Goal 11+. Tools*. 2016. Available from https://www.local2030.org/library/296/A-short-guide-to-Human-Settlements-Indicators-Goal-11.pdf

[6] Valencia SC, Simon D, Croese S, Nordqvist J, Oloko M, Sharma T, et al. Adapting the Sustainable Development Goals and the New Urban Agenda to the city level: Initial reflections from a comparative research project. *Int. J. Urban Sustain. Dev*. 2019; 11(1): 4-23. DOI: 10.1080/19463138.2019.1573172

[7] Sietchiping R, Reid J, Omwamba J. Implementing the SDGs and the New Urban Agenda. *Environment and Urbanization ASIA*. 2016; 7(2): x-xii. Doi:10.1177/0975425316660664.

[8] Pellicer S, Santa G, Bleda AL, Maestre R, Jara AJ, Skarmeta AG. A global perspective of smart cities: A survey. In: *Proc.7th. Int. Conf. on Innovative Mobile and Internet Services in Ubiquitous Computing* [Internet]; 2013 July 3-5; Taichung, Taiwan. IEEE, 2013; pp. 439-444. Available from https://ieeexplore.ieee.org/document/6603712.

[9] Fernandez-Anez, V. Stakeholders Approach to Smart Cities: A Survey on Smart City Definitions. In: Alba E, Chicano F, Luque G. editors. *Smart Cities. Smart-CT 2016. Lecture Notes in Computer Science*, vol 9704. Cham: Springer International Publishing, 2016. p. 157–67. https://doi.org/10.1007/978-3-319-39595-1_16.

[10] Gharaibeh A, Salahuddin MA, Hussini SJ, Khreishah A, Khalil I, Guizani M, Al-Fuqaha A: Smart cities: A survey on data management, security, and enabling technologies. *IEEE Communications Surveys & Tutorials*. 2017; 19(4): 2456-2501.

[11] Bibri SE, Krogstie J: Smart sustainable cities of the future: An extensive interdisciplinary literature review. *Sustain. Cities Soc*. 2017; 31: 83–212.

[12] Rajab H, Cinkel T: IoT based Smart Cities. In: *Proc. Int. Symp. on Networks, Computers and Communications (ISNCC)*; 2018 June 19-21; Rome, Italy. IEEE; pp. 1–4.

[13] Atzori L, Iera A, Morabito G. The internet of things: A survey. *Computer networks*. 2010 Oct 28;54(15):2787-805.
[14] Sundmaeker H, Guillemin P, Friess P, Woelfflé S. *Vision and challenges for realising the Internet of Things*. Cluster of European research projects on the internet of things, European Commission. 2010 Mar 3;3(3):34-6.
[15] Bélissent J. *Getting clever about smart cities: New opportunities require new business models*. Cambridge, Massachusetts, USA. 2010 Nov;193:244-77.
[16] Lee I, Lee K. The Internet of Things (IoT): Applications, investments, and challenges for enterprises. *Business horizons*. 2015 Jul 1;58(4):431-40.
[17] Gubbi J, Buyya R, Marusic S, Palaniswami M. Internet of Things (IoT): A vision, architectural elements, and future directions. *Future generation computer systems*. 2013 Sep 1;29(7):1645-60.
[18] Tanda A, De Marco A. Drivers of Public Demand of IoT-Enabled Smart City Services: A Regional Analysis. *J. Urban. Tech.* 2018 Oct 2;25(4):77-94.
[19] Knud Lasse Lueth. *State of the IoT 2018: Number of IoT devices now at 7B – Market accelerating* [Internet]. IoT Analytics; 2018 August 08. Available from https://iot-analytics.com/state-of-the-iot-update-q1-q2-2018-number-of-iot-devices-now-7b/
[20] Khatoun R, Zeadally S. Smart cities: concepts, architectures, research opportunities. *Communications of the ACM*. 2016 Jul 22;59(8):46-57.
[21] Zhang K, Ni J, Yang K, Liang X, Ren J, Shen XS. Security and privacy in smart city applications: Challenges and solutions. *IEEE Commun. Mag*. 2017 Jan 19;55(1):122-9.
[22] Martínez-Ballesté A, Pérez-Martínez PA, Solanas A. The pursuit of citizens' privacy: a privacy-aware smart city is possible. *IEEE Commun. Mag*. 2013 Jun 10;51(6):136-41.
[23] Elmaghraby AS, Losavio MM. Cyber security challenges in Smart Cities: Safety, security and privacy. *J. Advanced Research*. 2014 Jul 1;5(4):491-7.
[24] Roman R, Zhou J, Lopez J. On the features and challenges of security and privacy in distributed internet of things. *Computer Networks*. 2013 Jul 5;57(10):2266-79.
[25] Akhunzada A, Ahmed E, Gani A, Khan MK, Imran M, Guizani S. Securing software defined networks: taxonomy, requirements, and open issues. *IEEE Commun.Mag*. 2015 Apr 8;53(4):36-44.
[26] Zanella A, Bui N, Castellani A, Vangelista L, Zorzi M. Internet of things for smart cities. *IEEE Internet Things J*. 2014 Feb 14;1(1):22-32.
[27] Talari S, Shafie-Khah M, Siano P, Loia V, Tommasetti A, Catalão JP. A review of smart cities based on the internet of things concept. *Energies*. 2017 Apr;10(4):421.
[28] Silva BN, Khan M, Han K. Towards sustainable smart cities: A review of trends, architectures, components, and open challenges in smart cities. *Sustain. Cities and Society*. 2018 Apr 1;38:697-713.
[29] Khan M, Silva BN, Jung C, Han K. A context-aware smart home control system based on ZigBee sensor network. *KSII Transactions on Internet and Information Systems (TIIS)*. 2017;11(2):1057-69.
[30] Khan M, Silva BN, Han K. A web of things-based emerging sensor network architecture for smart control systems. *Sensors*. 2017 Feb;17(2):332.
[31] Khan M, Silva BN, Han K. Internet of things based energy aware smart home control system. *IEEE Access*. 2016 Oct 31;4:7556-66.
[32] Han J, Choi CS, Park WK, Lee I, Kim SH. Smart home energy management system including renewable energy based on ZigBee and PLC. *IEEE Trans. Consum. Electron*. 2014 Jul 14;60(2):198-202.
[33] Jabbar S, Khan M, Silva BN, Han K. A REST-based industrial web of things' framework for smart warehousing. *The J. Supercomputing*. 2018 Sep;74(9):4419-33.

[34] Kodali RK, Swamy G, Lakshmi B. An implementation of IoT for healthcare. 2015 *IEEE Recent Advances in Intelligent Computational Systems (RAICS)*. Trivandrum, Kerala, India. 2015.

[35] Solanas A, Patsakis C, Conti M, Vlachos IS, Ramos V, Falcone F, Postolache O, Pérez-Martínez PA, Di Pietro R, Perrea DN, Martinez-Balleste A. Smart health: A context-aware health paradigm within smart cities. *IEEE Commun. Mag*. 2014 Aug 7;52(8):74-81.

[36] Li X, Shu W, Li M, Huang HY, Luo PE, Wu MY. Performance evaluation of vehicle-based mobile sensor networks for traffic monitoring. *IEEE Trans. Vehicular Tech*. 2008 Sep 19;58(4):1647-53.

[37] Lee S, Yoon D, Ghosh A. Intelligent parking lot application using wireless sensor networks. In *2008 International Symposium on Collaborative Technologies and Systems* 2008 May 19 (pp. 48-57). IEEE.

[38] Al-Ali A, Aburukba R. Role of internet of things in the smart grid technology. *J. Comp. Commun.* 2015;3(05):229-33.

[39] Perera C, Jayaraman PP, Zaslavsky A, Georgakopoulos D, Christen P. Sensor discovery and configuration framework for the internet of things paradigm. In *2014 IEEE World Forum on Internet of Things (WF-IoT)* 2014 Mar 6 (pp. 94-99). IEEE.

Throughput Analysis of α-κ-μ Extreme/Gamma Composite Fading Channel

Rajesh Kumar[1], Poonam Yadav[1], Sandeep Kumar[2]
[1]Deptt. of ECE, NERIST, Nirjuli, 791109, AP, India
[2]CRL, Bharat Electronics Limited, Ghaziabad, 201010, UP, India
[1]itsrk2006@gmail.com, [1]poonam.yadav.25@gmail.com, [2]sann.kaushik@gmail.com

Abstract-In this paper, we consider α-κ-μ-Extreme/gamma composite fading channel model, which is used to characterize non-linear severe fading and shadowing impairments in enclosed wireless propagation scenarios. For this model, we first obtain the probability density function (PDF) expression for the received signal-to-noise ratio(SNR). Using the derived PDF, throughput analysis is provided by deriving expressions for the average channel capacity(ACC) and effective capacity(EC) in closed-form. The derived expressions are evaluated for several values of fading and shadowing indexes to study their effect on the ACC and EC. Furthermore, asymptotic analysis of ACC and EC are carried out to show better approximation under the high and low SNR regimes.

Keywords- Average channel capacity, enclosed propagation scenario, effective capacity.

I. Introduction

It is widely known that both multipath fading and shadowing have a major impact on wireless propagation links, resulting in the degradation of the transmitted signal. In genuine wireless fading channels, it is more precise to jointly express the shadowing and multipath fading features. To account for this composite fading phenomenon many statistical models are proposed in the literature [1] [2] [3] [4] [5].To characterise the performance of wireless communication system throughput analysis is exceptionally useful. Based on the Shannon channel capacity concept analytic results for the channel capacity over different composite fading models are investigated in several literatures. In [6], channel capacity analysis over κ-μ and η-μ channels with inverse gamma shadowing is provided. Performance of a Weibull/Lognormal fading is studied in [7]. Adding to this, some distinguished findings over the analysis of the throughput for different composite fading models can be found in [8] [9] [10]. Unfortunately, the Shannon channel capacity model lacks to account for the delay quality-of-service(QoS) guarantees of the next-generation wireless system scenarios such as IoT, device-to-device, body area networks, which supports real-time applications. This crucial limitation led to the framework of the EC [11], which has great potential to deal with QoS requirements of such networks. This concept has been widely exploited by researchers to analyse the performance of fading channels [12] [13] [14] [15]. EC analysis over the generalised lognormal(LN) fading channel is studied in [13] whereas [15] provides EC analysis over the Fischer-Snedecor F channel with diversity.

The generalised fading distribution such as α-κ-μ models consider the non-linearities of the wireless channel and correspond to the conditions of large number paths. Nevertheless, in the enclosed communication scenarios such as trains or airplanes where the number of paths is less leading to severe fading conditions, these models do not stand true. To address this in [16], α-κ-μ-Extreme fading distribution is formulated as a particular case of α-κ-μ distribution by taking κ → ∞ and μ → 0. Performance of multi-hop communication system based on decode-and-forward relaying technique is studied in [16] and [17] for the α-κ-μ-Extreme fading conditions. In the former channel capacity analysis is provided whereas later deals with the error performance analysis. In [18], performance of energy detector is studied over this distribution. Considering the gamma shadowing authors in [19] have presented envelope PDF for κ-μ-Extreme/gamma. In the view of the non-linearity of the propagation medium, α-κ-μ-Extreme/gamma is introduced in [2] and corresponding envelope PDF is derived. It is noticed from the extensive literature survey that; performance metric analysis is not performed over these channels in any technical literature so far. Inspired by this research gap, throughput analysis of the α-κ-μ-Extreme/gamma composite fading channel is presented in this work. To this end, the major contribution of this paper is summarized as follows: first, an expression for the SNR PDF of the proposed composite fading channel is derived. Then, utilizing it, expressions for ACC and EC of the proposed channel are obtained in closed-form. To better understand the limiting behaviour asymptotic analysis is also provided.

This paper is systemised as follows. Channel model is presented in section II along with the derivation of the SNR PDF for α-κ-μ-Extreme/gamma composite fading model. Section III contains the throughput analysis, which involves the formulation of the analytical expression for the channels' ACC and EC. Asymptotic analysis is covered in section IV. Section V presents numerical and simulation results. At the end section VI provides conclusions of the paper.

II. Channel Model

The α-κ-μ Extreme distribution appears as a significant particular case of the α-κ-μ distribution proposed in [21], by allowing $k \rightarrow \infty$ and $\mu \rightarrow 0$, the SNR PDF is given as [17]

$$f_\gamma(\gamma) = a\gamma^{\frac{\alpha}{4}-1}\exp\left(-b\gamma^{\frac{\alpha}{2}}\right)I_1\left(\sqrt{c\gamma^{\frac{\alpha}{2}}}\right) + \frac{\exp(-2m)}{2\sqrt{\gamma\bar{\gamma}}}\delta\left(\sqrt{\frac{\gamma}{\bar{\gamma}}}\right),\ \gamma>0 \tag{1}$$

Where

$$a = \frac{\alpha m\exp(-2m)}{\sqrt{\left(\bar{\gamma}\right)^{\alpha/2}}},\ b = \frac{2m}{\bar{\gamma}^{\alpha/2}},\ c = \frac{16m^2}{\left(\bar{\gamma}\right)^{\alpha/2}},\ m = \frac{\mu(\kappa+1)^2}{(2\kappa+1)}, \quad \text{and } \alpha,\kappa,\mu,>0$$

with α, κ, μ signifying the non-linearity, ratio of dominant power component to scattered wave power and multipath clustering, respectively. $\bar{\gamma}$ signifies average received SNR and $\delta(x)$ is Dirac delta function having value 1 at *x=0* and zero elsewhere. This fading model is used to specifically characterise the situation in wireless radio propagation through confined spaces [20] [21] [22].To model the shadowing phenomena LN and gamma distributions are widely described in the literature [23]. Here, we use gamma distribution to model the shadowing effect as it provides a simpler representation as compared to LN distribution. For the gamma distribution PDF is given as [24]

$$f_w(w) = \frac{w^{m_g}}{\bar{\gamma}^{m_g}\Gamma(m_g)} \exp\left(-\frac{w}{\bar{\gamma}}\right), \; m_g>0 \tag{2}$$

Where m_g is the shadowing parameter. Mathematically, SNR PDF of the composite fading distribution is represented as

$$f_\gamma(\gamma) = \int_0^\infty f_\gamma(\gamma/w) f_w(\text{w})dw \tag{3}$$

Where $f_\gamma(\gamma / w)$ is the conditional α-κ-μ-Extreme fading with average received SNR as w. The conditional PDF of instantaneous received SNR γ for α-κ-μ-Extreme fading distribution is obtained as

$$f_\gamma(\gamma/w) = \frac{\alpha m \exp(-2m)\gamma^{\frac{\alpha}{4}-1}}{w^{\frac{\alpha}{4}} \exp\left(2m(\gamma/w)^{\frac{\alpha}{2}}\right)} I_1\left(4m\sqrt{(\gamma/w)^{\frac{\alpha}{2}}}\right) + \frac{\exp(-2m)}{2(\gamma w)^{0.5}}\delta\left(\sqrt{\frac{\gamma}{w}}\right) \tag{4}$$

Keeping (2) and (4) in (3) and after a few steps of mathematical reduction we get the SNR PDF for the shadowed α-κ-μ-Extreme/gamma fading channel as

$$f_\gamma(\gamma) = \Re\gamma^{m_g-1}\sum_{i=1}^{N} w_i a_i \exp(-b_i\gamma) \tag{5}$$

Where $\Re = \frac{(2m)^{\frac{2m_g}{\alpha}+\frac{1}{2}}}{\bar{\gamma}^{m_g}\Gamma(m_g)\exp(2\text{m})}$; $b_i = \frac{1}{\bar{\gamma}}\left(\frac{2m}{x_i}\right)^{\frac{2}{\alpha}}$; $a_i = x_i^{\left(-\frac{1}{2}-\frac{2m_g}{\alpha}\right)} I_1\left(\sqrt{8mx_i}\right)$. w_i is the weight, a_i is the abscissas and N is number of terms in Gaussian-Laguerre summation.

III. Throughput Analysis

A. Average channel capacity

In this section we obtain the mathematical formulation of the ACC over the α-κ-μ-Extreme/gamma fading channel. Mathematically, ACC is expressed as

$$\begin{aligned}\bar{C} &= \text{E}[\log_2(1+\gamma)] \\ &= \int_0^\infty \log_2(1+\gamma) f_\gamma(\gamma)d\gamma\end{aligned} \tag{6}$$

Substituting (5) in (6) and using the Meijer G expansion of the exponential and logarithmic function using equation 11 of [25] we get

$$\bar{C} = \frac{\Re}{\ln(2)}\sum_{i=1}^{N} w_i a_i \int_0^\infty \gamma^{m_g-1} G_{22}^{12}\left(\gamma \,\middle|_{10}^{11}\right) G_{01}^{10}\left(b_i\gamma \,\middle|_{0}\right) d\gamma \tag{7}$$

This integral is solved using equation 07.34.21.0011.01 of [26] and the analytical expression for the ACC of the mentioned composite fading channel is obtained as

$$\bar{C} = \frac{\Re}{\ln(2)}\sum_{i=1}^{N} w_i a_i G_{23}^{31}\left(b_i \,\middle|_{0, \; -m_g, -m_g}^{-m_g, 1-m_g}\right) \tag{8}$$

B. Effective capacity

The consequences of statistical delay QoS on the communication system can indeed be characterise using EC concept. The normalised EC over the fading channel is given in [27] as

$$R = -\frac{1}{A}\log_2\left(\mathrm{E}\left[(1+\gamma)^{-A}\right]\right) \tag{9}$$

$$= -\frac{1}{A}\log_2\left(\int_0^\infty (1+\gamma)^{-A} f_\gamma(\gamma)d\gamma\right) \tag{10}$$

Where A stands for the delay constraint parameter and is defined in [27]. Substituting (5) in (10) and using equation 8.4.2.5 of [28] we get

$$R = -\frac{1}{A}\log_2\left(\Re\sum_{i=1}^{N} w_i a_i \int_0^\infty \frac{1}{\Gamma(A)}\gamma^{m_g-1}\exp(b_i\gamma)G_{11}^{11}\left[\gamma\,\Big|_{0}^{1-A}\right]d\gamma\right) \tag{11}$$

Transforming the exponential term using equation 11 of [25] into Meijer G function then employing equation 07.34.21.0011.01 of [26], the integral in (11) can be solved to obtain the expression of EC over the proposed composite fading channel as

$$R = -\frac{1}{A}\log_2\left(\Re\sum_{i=1}^{N} w_i a_i \frac{b_i^{-m_g}}{\Gamma(A)} G_{21}^{12}\left[\frac{1}{b_i}\,\Big|_{0}^{1-A,1-m_g}\right]\right) \tag{12}$$

IV. Asymptotic Analysis

The asymptotic analysis allows us to understand the performance of the system in practical scenarios and also determines how far from the asymptotically achievable performance a practical system is. So, to understand the behaviour of the system under the limited regime, in this section asymptotic analysis of the throughput performance parameter is provided.

A. Asymptotic analysis of ACC

The asymptotic capacity can be obtained at high SNR regime by considering $\gamma \to \infty$, then $\log_2(1+\gamma) \simeq \log_2(\gamma)$ .Then the ACC expression can be given as

$$\overline{C}^{\infty} = \frac{1}{\ln(2)}\int_0^\infty \ln(\gamma) f_\gamma(\gamma)d\gamma \tag{13}$$

$$= \frac{\Re}{\ln(2)}\sum_{i=1}^{N} w_i a_i \int_0^\infty \ln(\gamma)\gamma^{m_g-1}\exp(-b_i\gamma)d\gamma \tag{14}$$

Using equation 4.352.1 of [29] the integral in (14) can be solved to obtain the asymptotic ACC under high SNR regime as

$$\overline{C}^{\infty} = \frac{\Re}{\ln(2)}\sum_{i=1}^{N} w_i a_i \frac{\Gamma(m_g)}{b_i^{m_g}}\left[\Psi(m_g) - \ln(b_i)\right] \tag{15}$$

Where $\Psi(.)$ is the psi function defined by equation 8.360.1 of [29].

B. Asymptotic Analysis of EC

By setting $\gamma \to \infty$ in (9) we get the high SNR approximation expression for the EC as

$$R^{\infty} = -\frac{1}{A}\log_2\left(\mathrm{E}\left[(\gamma)^{-A}\right]\right) \tag{14}$$

By invoking the definition of the ith moment of γ, delineated as,

$$E\left[\gamma^{i}\right] = \int_{0}^{\infty} \gamma^{i} f_{\gamma}(\gamma) d\gamma \tag{15}$$

Then, with the aid of (5) and equation 3.326.2 of [29], (14) can be further solved to get the EC expression for α-κ-μ- Extreme/gamma fading channel in high SNR area as

$$R^{\infty} = -\frac{1}{A}\log_2\left(\Re\sum_{i=1}^{N} w_i a_i \frac{\Gamma\left(m_g - A\right)}{b_i\left(m_g - A\right)}\right) \tag{16}$$

Low SNR analysis using γ, leads to inaccurate results so when the system is operating under low SNR conditions which happens to be specifically in the cases of mobile networks, EC is analysed in the terms of normalised energy transmitted per information bit E_b/N_0 as presented in [30]

$$R\left(\frac{E_b}{N_0}\right) \approx S_0 \log_2\left(\frac{E_b/N_0}{E_b/N_{0\min}}\right) \tag{17}$$

where the parameters E_b/N_{0min} and S_0 indicate the low SNR nature essential for transmission of non-negative rate and spectral efficiency curve slope to ensure non-zero power level, respectively. $R'(.)$ and $R''(.)$ are 1st and 2nd order derivatives of EC derived from (9) and are used to generate the following two parameters:

$$\frac{E_b}{N_{0\min}} = \frac{1}{R'(0,\theta)}; S_0 = \frac{-2\ln(2)\left(R'(0,\theta)\right)^2}{R''(0,\theta)} \tag{18}$$

$$R'(0,\theta) = \frac{E[\gamma]}{\ln 2}; R''(0,\theta) = \frac{-(A+1)E[\gamma^2]}{\ln 2} + \frac{A\left(E[\gamma]\right)^2}{\ln 2} \tag{19}$$

The E[.] can be evaluated using (15). Substituting (5) in (15) and using equation 3.326.2 of [29], considering *i*=-1 and *i*=-2, the first and second moment can be evaluated as

$$E[\gamma] = \Re\sum_{i=1}^{N} w_i a_i \frac{\Gamma\left(m_g + 1\right)}{b_i\left(m_g + 1\right)} \tag{20}$$

$$E[\gamma^2] = \Re\sum_{i=1}^{N} w_i a_i \frac{\Gamma\left(m_g + 2\right)}{b_i\left(m_g + 2\right)} \tag{21}$$

To generate the expression for EC in the low SNR conditions, (20) and (21) are substituted in (19) and then using (18), expression for E_b/N_{0min} and S_0 is evaluated. Thereafter, substituting in (16), the generalised expression under low SNR is obtained as

$$R\left(\frac{E_b}{N_0}\right) \approx \frac{2\log_2\left(\frac{\left(\Re\sum_{i=1}^{N} w_i a_i\,\Gamma\left(m_g+1\right)/b_i\left(m_g+1\right)\right)E_b/N_0}{\ln 2}\right)}{\left(\frac{(A+1)\Re\sum_{i=1}^{N} w_i a_i\,\Gamma\left(m_g+2\right)/b_i\left(m_g+2\right)}{\left(\Re\sum_{i=1}^{N} w_i a_i\,\Gamma\left(m_g+1\right)/b_i\left(m_g+1\right)\right)^2}-A\right)} \quad (22)$$

V. Results and Discussion

This section elaborates the numerical analysis findings for several shadowing and multipath fading scenarios using the aforementioned theoretical formulations. MATLAB2015a was used to simulate the analytical findings. Based on (8), figure1 illustrates the ACC as a function of $\bar{\gamma}$ over α-κ-μ-Extreme/gamma composite fading channels. It is observed that there is a substantial improvement in the ACC as the fading parameters α and m increments. It also depicts that as the shadowing severity reduces that is the value of m_g is increased ACC improves significantly.

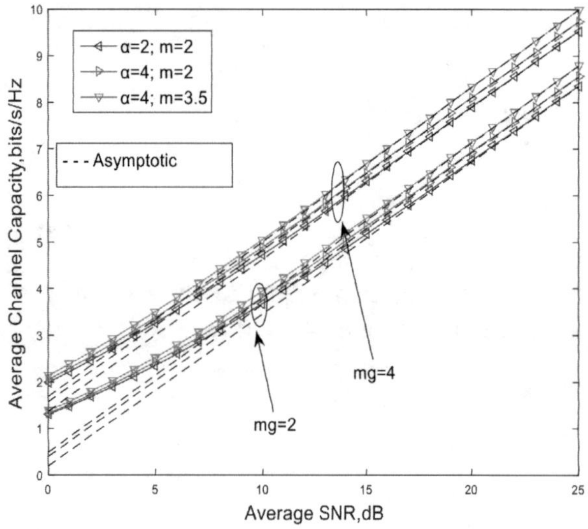

Figure 1 Impact of fading parameters on ACC

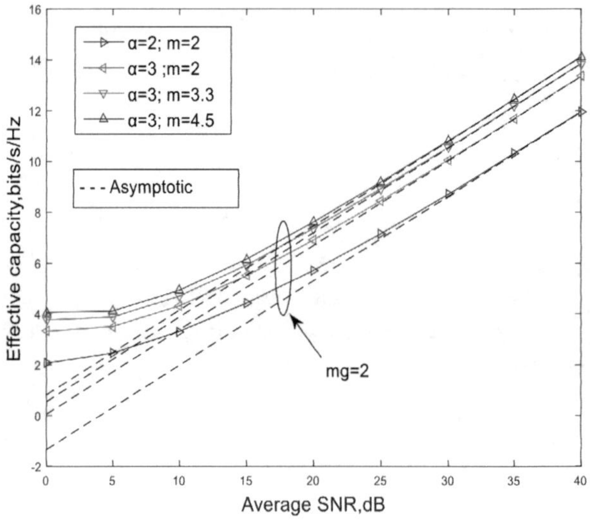

Figure 2 Impact of multipath fading parameters on EC

At α=4 and m=3.5 considering $\bar{\gamma}$ =10 dB as shadowing parameter value is doubled ACC improves by 25%. At m_g =2, m=2, and $\bar{\gamma}$ =10 dB as α increases from 2 to 4 there is a 5.2% improvement in the ACC. From this, it can be depicted that at the same $\bar{\gamma}$ value shadowing parameter is more influential than the fading parameters. Furthermore, the obtained curves for the channel are exceptionally tight with the high approximation curves, which become more exact as the average SNR increases. Similar behaviour is observed for the EC from figure 2. From figure 3 quantitatively it can be stated that at $\bar{\gamma}$ =5 dB, as shadowing parameter m_g rises from 2 to 4 the EC is increased by 40%. Also, at m_g =4 as $\bar{\gamma}$ increases to 10 dB from 20 dB EC improves by 77%. From figure 4 it can be noticed that EC consistently reduces over the larger delay exponent θ, which implies that delay constraint becomes larger less EC system can handle. For instance, at θ=5, as T increases from 1 second to 4 seconds, EC deteriorates by 64%. Furthermore, when the equivalent block length T increases, the space between the curves narrows, implying that the longer T, the more pronounced the effect.

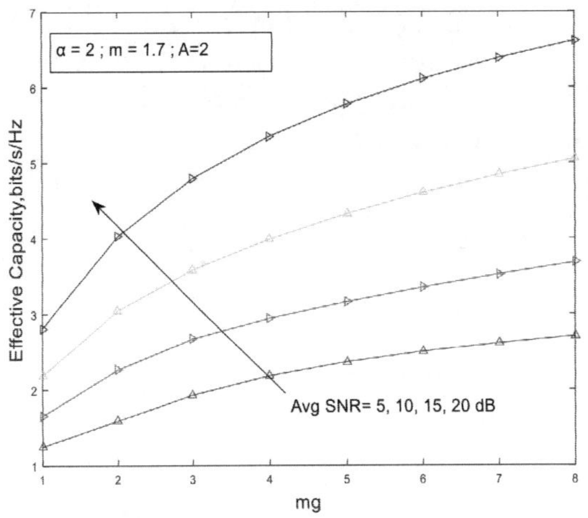

Figure 3 EC vs m_g for different values of $\bar{\gamma}$

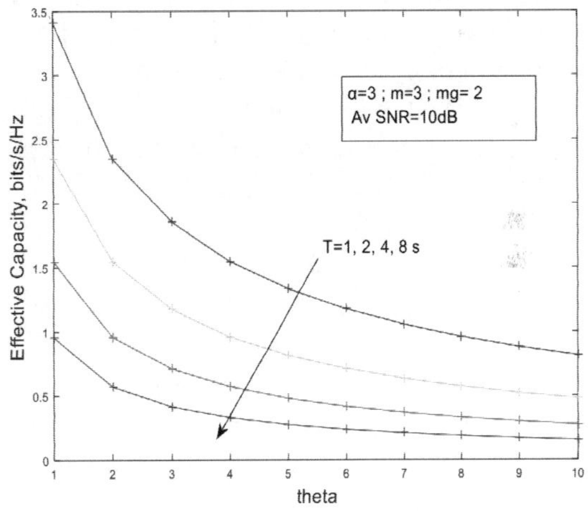

Figure 4 EC vs θ for distinct values of T

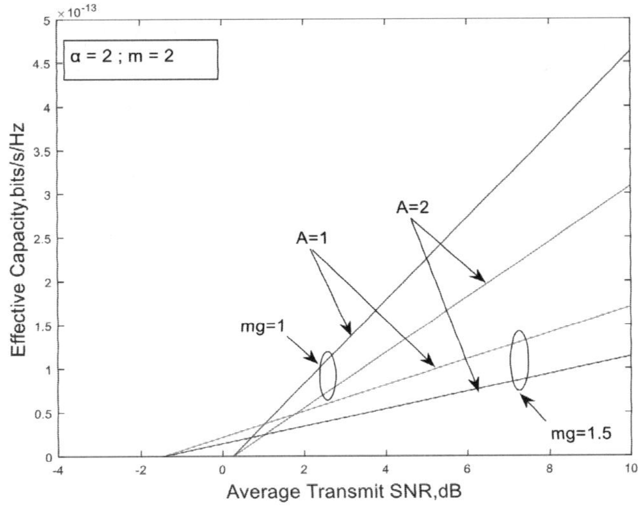

Figure 5 EC versus E_b/N_0 under low SNR regime

Figure 5 illustrates the EC over the low E_b/N_0 approximation. The nature of the plot depicts that as the delay constraint A is increased, EC degrades whereas E_b/N_0 remains unaffected. However, with increment in the shadowing factor the E_b/N_0 falls as it can be observed that as m_g increases by 33%, E_b/N_0 falls from 0.4 dB to -1.6 dB.

VI. Conclusion

By obtaining the general and explicit analytical formulation of ACC and EC, we investigated the system throughput of α-κ-μ-Extreme/gamma composite fading channel in this study. The expressions are derived in the form of the Meijer G function which makes it tractable for the analysis. The obtained results imply that various physical channel characteristics have an impact on throughput performance. The behaviour of the channel at the limiting conditions is revealed by obtaining simple asymptotic expression for throughput under high and low SNR regimes. This work can be further extended to analyse other performance metrics such as ABER/ASEP of various wireless communication scenarios also performance of the diversity techniques can be studied over the proposed composite fading channel.

References

[1] P. Yadav, S. Kumar, R. Kumar, "A Review of Transmission Rate over Wireless Fading Channels: Classifications, Applications, and Challenges" Wireless Pers Commun, vol. 22, Sep 2021, pp. 1709-1765. https://doi.org/10.1007/s11277-021-08968-1.

[2] P. C. Sofotasio and S. Freear, "The α-κ-μ/gamma distribution: A generalized non-linear multipath/shadowing fading model," in *Annual IEEE India Conference*, Hyderabad, India, 2011.

[3] H.S. Al-Raweshidy and H. Al- Hamood, "Unified modeling of composite κ-μ/gamma, η-μ/gamma, and α-μ/gamma fading channels using a mixture gamma distribution with applications to energy detection," *IEEE Antennas Wireless Propag. Lett,* vol. 16, pp. 104-108, Apr 2017.

[4] P. S. Bithas, "Weibull-gamma composite distribution: alternative multipath/shadowing fading model," *Electronics Letters,* vol. 45, no. 14, pp. 749-751, Jul 2009.

[5] M. Kaur,V. Kansal, and S. Singh., "A Review Paper on Composite Fading Models," *Compliance Engineering Journal,* vol. 10, no. 9, pp. 354-357, 2019.

[6] D. Pant et al., "Channel Capacity Analysis of Wireless System under ORA scheme over κ-μ/Inverse Gamma and $\eta-\mu/$ Inverse Gamma Composite Fading Models," in *International Conference on Electrical and Electronics Engineering (ICE3)*, Gorakhpur, 2020.

[7] R. Singh et. al., "A New Approximate Closed-Form Distribution and Performance Analysis of a Composite Weibull/Log-Normal Fading Channel," *Wireless Pers Commun,* vol. 92, pp. 883-900, 2017.

[8] O. S. Badarneh et al., "Performance Analysis of Mobile IoT Networks over Composite Fading Channels," in *International Wireless Communications and Mobile Computing (IWCMC)*, Limassol,Cyprus, 2020.

[9] P. C. Sofotasios et al., "The kappa-mu /IG Composite Statistical Distribution in RF and FSO Wireless Channels," in *IEEE 78th Vehicular Technology Conference (VTC Fall)*, Las Vegas, NV, 2013.

[10] S.Kumar et al., "Performance analysis of Hoyt-lognormal composite fading channel," in *International Conference on Wireless Communications, Signal Processing and Networking (WiSPNET)*, Chennai, 2017.

[11] D. Wung and R.Negi, "Effective capacity: A wireless link model for support of quality of service," *IEEE Trans. Wireless Commun.,* vol. 2, no. 4, pp. 630-643, 2003.

[12] H. Al-Hmood and H. Al-Raweshidy, "Unified approaches based effective capacity analysis over composite α−η−μ/gamma fading channels," *Electronics Letters* , vol. 54, no. 13, pp. 852 - 853, June 2018 .

[13] P. Yadav, R. Kumar and S. Kumar, "Effective capacity analysis over generalized lognormal shadowed composite fading channels," *Internet Technology Letters,* vol. 3, no. 5, pp. e171, 2020.

[14] P. Yadav, S. Kumar and R. Kumar, "Effective Capacity Analysis over α-κ-μ/Gamma Composite Fading Channel," in *2nd International Conference on Advances in Computing, Communication Control and Networking (ICACCCN)*, Greater Noida, India, 2020.

[15] M. Kaur and R. K. Yadav, "Effective Capacity Analysis Over Fisher-Snedecor F Fading Channels with MRC Reception," *Wireless Personal Communication,* vol. 121, no. 4, 2021.

[16] T. R. Rasethuntsa et al., "On the Outage Probability and BER of a DF-Based Multi-Hop System Over α–κ–μ and α–κ–μ-Extreme Fading Channels," *AEU - International Journal of Elect and Commun,*vol.124,Sep 2020 doi: 10.1016/j.aeue.2020.153324.

[17] T.R. Rasethuntsa et al., "A Comprehensive Performance Evaluation of a DF-Based Multi-Hop System Over α-κ-μ and α-κ-μ/Extreme Fading Channels," *arXiv preprint arXiv:1903.09353* (2019).

[18] T. R. Rasethuntsa and S. Kumar, "An integrated performance evaluation of ED-based spectrum sensing over α-κ-μ and α-κ-μ/Extreme fading channels," *Trans on Emerging Telecommun Technologies,* vol. 30, no. 5, p. e3569, 2019.

[19] P. C. Sofotasios and S. Freear, "The κ-μ Extreme/Gamma Distribution: A Physical Composite Fading Model," in *IEEE WCNC*, Cancun, Mexico, 2011.

[20] L. Bakir and J. Frolik, "Diversity gains in two-ray fading channels," *IEEE Trans. Wireless Commun.,* vol. 8, no. 2, pp. 968-977, 2009.

[21] G. D. Durgin et al., "New analytical models and probability density functions for fading in wireless communications," *IEEE Trans Commun.,* vol. 50, no. 6, pp. 1005-1015, 2002.

[22] J. Frolik, "On appropriate models for characterizing hyper-Rayleigh fading," *IEEE Trans. on Wir. Com.,* vol. 7, no. 12, pp. 5202-5207, 2008.

[23] P. M. Shankar, Fading and shadowing in wireless systems, 1st ed., NY: Springer, 2012.

[24] M. K. Simon and M. S. Alouini, Digital Communications over Fading Channels: A Unified Approach to Performance Analysis, 2nd ed., NY: John Wiley & Sons, 2004.

[25] V. S. Adamchik and O. I. Marichev, "The algorithm for calculating integrals of hypergeometric type functions and its realization in reduce system," *Proceedings of the international symposium on Symbolic and algebraic computation,* pp. 212-224, 1990.

[26] Available: http://functions.wolfram.com/PDF/MeijerG.pdf.Accessed on Nov 2021.

[27] S. Chen et al., "Effective Rate of MISO systems over Fisher-Snedecor F fading channels," *IEEE Communications Letters,* vol. 22, no. 12 , pp. 2619 - 2622, Dec 2018.

[28] A.P. Prudnikov et al., Integrals and series: more special functions, 3rd ed., New York: Gordon & Breach Sci. Publ., 1990.

[29] I. S. Gradshteyn and I. Ryzhik, Table of Integrals,Series, and Products(8th ed), Elseiver, 2014.

[30] A. Lozano et al. "Multiple-antenna capacity in the low-power regime," *IEEE Trans. Inf. Theory,* vol. 49, no. 10, pp. 2527–2544, 2003.

Journal of Physics: Conference Series **2236** (2022) 012010 doi:10.1088/1742-6596/2236/1/012010

Printed monopole antenna design with CPW fed for ultra wideband application

Samom Jayananda Singh [1,3] ,Rajesh Kumar[1] and M M Dixit[2]

[1]Dept of Electronics and Communication Engg, NERIST, Nirjuli Arunachal Pradesh India ,791109

[2]Dept of Mathematics, NERIST, Nirjuli Arunachal Pradesh India, 791109

[3]jayanandmetal.inc@gmail.com

Abstract. A circular monopole antenna fed by co-planar waveguide(CPW) for ultra wideband application is described in this article. The usage of a co-planar ground plane help to increase bandwidth for a frequency range 5G frequency spectrum's mid band. The proposed antenna is designed to operate from 4.8GHz to 14.10GHz. With this proposed antenna, a suitable omnidirectional radiation pattern can be created. Comparison of fractional bandwidth value for various gap between the ground plane and the patch for satisfying the criteria for UWB has been analysed here.

1. Introduction

The development of ultra wideband systems is progressing at a breakneck pace. Many microwave techniques use ultra-short pulses in the nanosecond range to operate on the UWB band. FCC committee designated the frequency range from 3.1 and 10.6 GHz with resonance frequency of 7.5GHz bandgap of at least 500 MHz, UWB systems have gained attention. One of the major benefits of UWB is that it has the capable of holding strong data rate minimum utilization of power, easy hardware layout in real applications, and have a fair multipath resistance. UWB technology is generally used in a various way of applications, for having a strong data transmission which is meant for indoor communication purposes, especially for short-range transmission[1].For ultra wideband (UWB) operation and omnidirectional radiation characteristics, printed monopole antennas (PMA) are the best option. UWB refers to a total frequency range of above 500 MHz or a percentage frequency range of more than 20%. To accomplish the working of UWB within the frequency range from 8 GHz to 18 GHz, monopole antenna having a rectangular patch with a magnetodielectric material cover layer is used[2].When compared to wire antennas, planar antennas with parasitic patches perform better in emitting the Electric fields and Magnetic field by lowering the antenna's physical height[3]. Microstrip-fed printed planar UWB monopole antenna that is suited for UWB operation. Single, dual, and triple band rejection antennas can readily be added for wide matching, strong omnidirectional coverage, and consistent gain indicate that the developed antenna is suitable for a variety of UWB applications[4]. Printed antennas of various shaped were started for enhancement of resistive and reactive impedance components of the patch. In UWB technology, a two-sided PMA having a minor footprint and allowable range of impedance properties is observed. Antenna characteristics are optimised, and the ground plane is modified by cutting slots on the top edge to generate a regular sawtooth shape to increase impedance bandwidth. With the advancement of modern etching method, there has been a significant expansion in the field of wireless technology[4,5].In[6],the antenna is a small circular disc monopole UWB antenna is meant for antenna's input impedance matching by

Journal of Physics: Conference Series **2236** (2022) 012010 doi:10.1088/1742-6596/2236/1/012010

tapering the microstrip feeding part with ground plane. It gives radiates a uniform electric fields which covers in all directions and maintaining a good gain in comparison with a normal wired antenna. In [7],antennas with rectangular patch having multiple feeds line are employed to increase the impedance bandwidth. Wideband antenna characteristics result from the effects of patch width fluctuation and associated offset feed. Off-centered feed Printed Rectangular Monopole Antennas enhance bandwidth with greater performance than single-center-feed PRMA. With a bandwidth of 8.4 GHz and a reflection co-efficient of below -10dB, an antenna with dual steps and a circular slot is envisioned. Advance band-notched with appropriate cutting part of the patch, promote a electric field scenario[8].The use of bevels in the patch and ground plane of a printed monopole antenna to increase bandwidth in the frequency range of 3 to 15 GHz is demonstrated. By adjusting the dimensions of the bevels, the bandwidth can be enhanced even further [9].In order to broaden the bandwidth, a defective ground structure design with a CPW supplied line in the printed monopole antenna is introduced. A defective ground plane causes an uneven current distribution in the ground plane. This uneven current distribution causes capacitance and inductance to vary their properties[10]. In,[11] circular polarized antenna, quadrilateral patch structure in a CPW fed planar monopole antenna designed is employed. A proper maintain of impedance by observing the reactive and resistive impedance component bandwidth augmentation of the bandwidth can be make possible, considering the directivity of the antenna is carried out correctly. Elliptical patch with various ground plane designs such as rectangle, modified half circular, and sleeves is offered. This strategy aids in the attainment of a suitable bandwidth as well as the improvement of S-parameter features[12].A printed monopole antenna with a half-elliptical patch form has been introduced to increase bandwidth upto 12.5 GHz. Cutting sectors slots on the patch can be used to increase bandwidth even more. It then makes an arc-shaped patch to act as a filter[13].In[14], a printed monopole antenna with a 9-shaped patch structure that is well suited for 4G and 5G technologies. Truncation in the ground plane with a fixed strip line is planned for better impedance matching. A suitable impedance bandwidth can be maintained by changing the feeding arrangement of the antenna with a Psi structure patch. The bandwidth can be increased to some extent by cutting slots in the patch and ground plane[15].

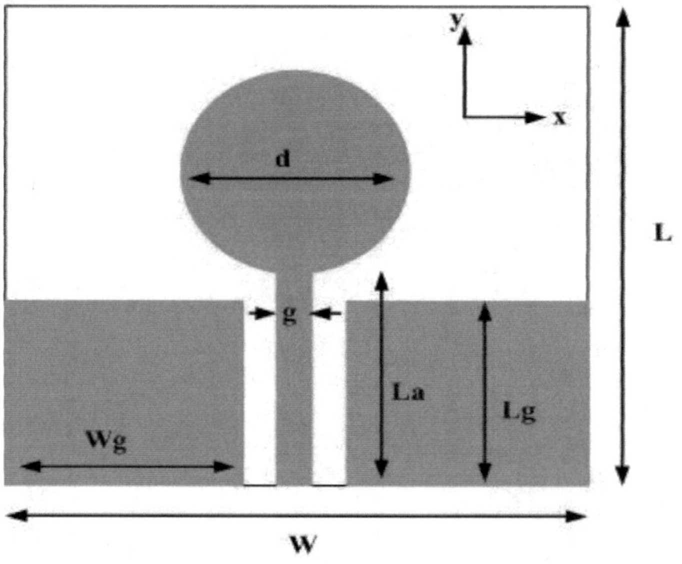

Figure 1. Geometry of proposed antenna.

2. Antenna design

The design for circular disc PMA for single element is similar to circular cavity waveguide modes which has dominant mode of TM_{11} (Transverse Magnetic).The resonant frequency of the printed antenna, for circular patch shape is given by[10].

$$f_{res} = \frac{\chi_{nm} c}{2\pi r \sqrt{\varepsilon_{rr}}} \tag{1}$$

$$f_{res} = \frac{1.8412c}{2\pi r \sqrt{\varepsilon_{rr}}} \tag{2}$$

Where,

f_{res} = resonant frequency of the printed antenna

r = radius of the circular patch .

ε_{rr} = permittivity of the dielectric substrate.

c = velocity of light for free space.

χ_{nm} = the mth root of the derivative of the Bessel function of order 'n'

The wavelength(λ_g) was used to determine the design parameters, which relates to a lower cut-off frequency of 3.1GHz[15]. Equation 3, provides the diameter of the circular disc patch(d), feed(L_a) and feed gap as given below.

$$\frac{\lambda_g}{8} \le d \le \frac{\lambda_g}{4} \tag{3}$$

The diameter of the circular patch(d) and the length of the feed are added together to get the radiator's effective length (La).

$$(d + L_a) \approx \frac{\lambda_g}{3} \tag{4}$$

The feed gap is the difference between the feed length (La) and the maximum coplanar ground plane length (Lg).Equation 5, can be used to get the appropriate feed gap value.

$$0.01\lambda_g \le (L_a - L_g) \le 0.02\lambda_g \tag{5}$$

The dimensions of the dielectric substrate (Ls,Ws, and hs) of the FR-4 epoxy dielectric substrate are chosen to satisfy the following equation from 3 to 8.

.

$$\frac{\lambda_g}{4} \le Ls \le \frac{\lambda_g}{2} \tag{6}$$

$$Ws = 2d \tag{7}$$

$$hs = 0.01\lambda_g \tag{8}$$

In order to satisfied the condition of UWB, fractional bandwidth plays a major role to met this condition. The value of fractional bandwidth must be greater than 0.2 or 20% for UWB .

Fractional bandwidth formula is given by

$$FBW = \frac{2(f_{high} - f_{low})}{(f_{high} + f_{low})} \tag{9}$$

Where,

f_{high} = upper 10dB cut-off frequency.
f_{low} = lower 10dB cut-off frequency.

Table 1. Proposed antenna parameters.

Designed antenna parameters	Dimension value(mm)
Substrate Length,Ls	30
Substrate Width, Ws	35
Substrate thickness, hs	0.8
Diameter of circular patch ,d	15
Ground Length,Lg	15.9
Ground Width,Wg	30
Feed Length, La	17.54
Feed width, g	1.6

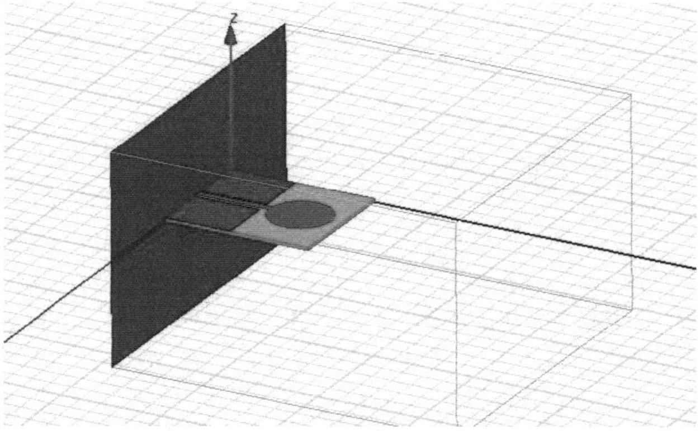

Figure 2. 3D Design of Simulated Proposed antenna.

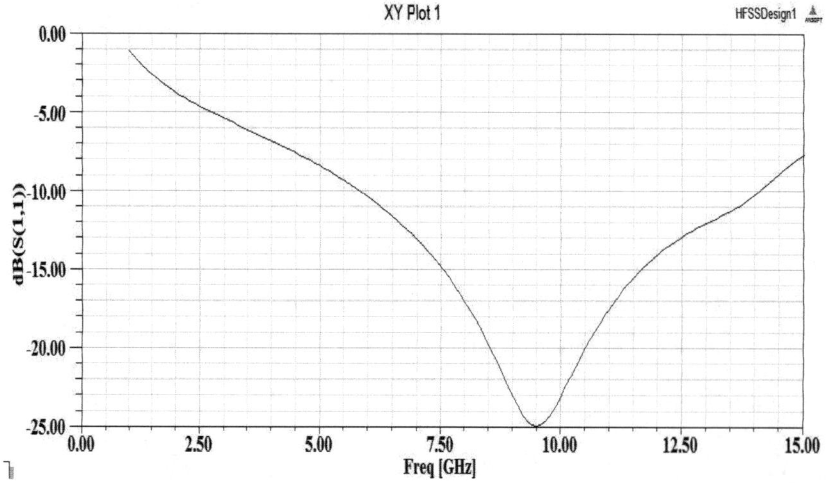

Figure 3. Reflection coefficient(S_{11}) vs frequency plot.

The design relationship is used to compute the various parameters of the proposed antenna [10][12][15]. The intended single element has a reflection co-efficient of -24.52dB at 9.50 GHz. In the frequency spectrum 5.8GHz to 14.10GHz depicted in Fig.3, the reflection co-efficient S_{11} is less than -10dB. The developed antenna's voltage standing wave ratio (VSWR) has produced fair outcome within the frequency range from 5.8GHz to 14.10GHz, The designed achieved a VSWR of 1.55 (which is less than 2)at the central frequency of 9.50 GHz, which is best suited for UWB .

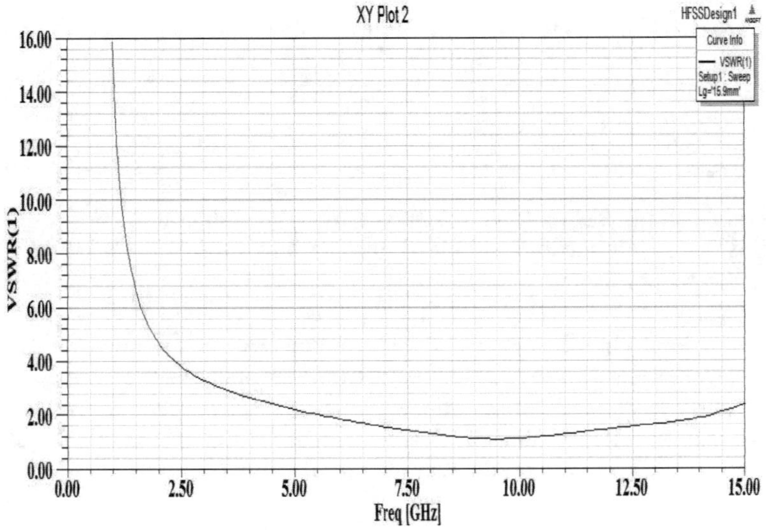

Figure 4.VSWR vs frequency plot

Figure 4 shows the VSWR plot, which describes the importance of input impedance of the antenna that is responsible for radiation when the reactive value of impedance is reduced to minimum and resistive part of the impedance is equal to 50Ω which is the characteristic impedance of co-axial cable. The VSWR of the planned antenna indicates that the input port impedance is adequately matched. Gap between the path and ground plane of CPW fed plays a major role in extending the bandwidth from the previous design of circular disc PMA. The best performance regarding the bandwidth plot is observed at the resonating frequency of 9.50 GHz. Lower cut off frequency is at 5.81GHz and upper cut off frequency is at 14.01GHz,and providing a bandwidth of 8.21GHz,and the calculation of fractional bandwidth value of 0.87,which is best suited of UWB.

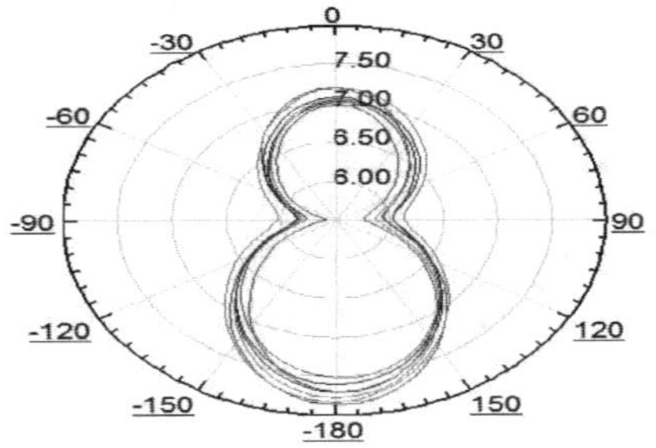

Figure 5. Radiation pattern of E plane.

In the E plane, a proper omnidirection radiation pattern can be seen. The graph shows a better performance in terms of the proposed antenna's radiation pattern, as shown in Fig 5.

3. Result and Discussion

Fractional bandwidth of about 87.36% has been observed in this proposed antenna. It gives a good significant in the bandwidth enhancement.The designed antenna cover the bandwidth from (5.81GHz-14.01GHz) provided VSWR $\leq$ 2.Besides bandwidth improvement, a fair radiation pattern of omnidirectional can be seen with this proposed antenna. A well matched impedance is observed for creation of proper radiation pattern is observed with this antenna design.

4. Conclusion

Printed monopole antenna with CPW fed in the mid frequency band of 5G is discussed in this study. Metamaterial and superstrate should be employed in the design to increase gain. Modification in the circular disc by cutting the edge will help in bandwidth. The suggested antenna will be appropriate for upcoming low-distance wireless communications

References

[1] Elfergani IT,Rodriguez J,Otung I,Mshwat W, Abd-Alhameed RA 2018 Slotted Printed Monopole UWB Antennas with Tunable RejectionBands for WLAN /WiMAX and X -bAND Coexistence,Radio Engineering.27(3):694-702.

[2] R S Kshetrimayum and R Bhattacharjee 2015 Bandwidth enhancement of printed monopole antennas using magnetodielectric cover ,IEEE AppliedElectromagnetics Conference (AEMC).

[3] K Ayyadurai ,G M Muthu 2015 Bandwidth enhancement of capacitive fed monopole antenna using parasitic patches".International Journal of Microwave and Wireless Technologies.

[4] N A Murugan,R Balasubramanian , H R Patnam 2018Printed Planar Monopole Antenna Design for Ultra-Wideband Communication",Radio electronics and communication system Vol61.No3.

[5] R Azim,M T Islam , N Misran 2011 Ground modified double-sided printed compact UWB antenna",Electronics letters,Vol.47 No.1.

[6] Zhang.Y,Nakata T,Miyashita T 2018 A minature cirular disc monopole UWB antenna with a tapered feed line and a circular ground",In: Proceedings of the IEEE China-japan joint microwave,shanghai,china ,2008,pp.441-414,ISBN:978-1 4244-3821.

[7] K P Ray, S S Thakur , S S Kakatkar 2014 Bandwidth Enhancement Techniques for Printed Rectangular Monopole Antenna". IETE Journal of Research, 60(3), 249–256.

[8] O Ahmed ,A.R.Sebak 2008 A Printed Monopole Antenna With Two Steps and a Circular Slot for UWB Applications". IEEE Antennas and Wireless Propagation Letters, 7, 411–413..

[9] P S Bakariya , SDwari ,M.Sarkar 2015 Triple band notch UWB printed monopole antenna with enhanced bandwidth,Int. J. Electron. Commun. (AEÜ) 69 pp 26–30.

[10] A.Singh and S.Singh 2015 A novel CPW-fed wideband printed monopole antenna with DGS, Int. J. Electron. Commun. (AEU) 69 (2015) 299–306.

[11] P. Chaudhary and A. Kumar 2019 Compact ultra-wideband circularly polarized CPW-fed monopole antenna" Int. J. Electron. Commun. (AEÜ) 107 pp137–145.

[12] A A Adam , S K ARahim ,K G Tan, A W Reza 2013 Design of 3.1–12 GHz Printed Elliptical Disc Monopole Antenna with Half Circular Modified Ground Plane for UWB Application" Wireless Pers Commun 69:pp535–549.

[13] R. Eshtiaghi, R. Zaker, J Nouronia , C Ghobadi 2010 UWB semi-elliptical printed monopole antenna with sub band rejection filter" Int.J.Electron.Commun.(AEÜ)64pp133–141

[14] A Pandey , R Mirsa 2016Compact Dual Band Monopole Antenna for RFID and WLAN Applications " International Conference on Processing of Materials, Minerals and Energy.

[15] A S Kundu 2019 Experiment study of a printed ultra -wideband modified circular monopole antenna ,Microwave Optical Letter.

Formant Measure of Indian English Vowels for Speaker Identity

Vinod Kumar[1], Om Prakash Roy[2]

[1, 2] Department of Electrical Engineering, NERIST, Nirjuli, Arunachal Pradesh
E-mails: {vinodnerist, oproy61} @ gmail.com

Abstract. With the growth of communication technology, reliability and security of communication devices became a challenging job. Voice biometrics are becoming increasingly popular as a promising alternative to traditional passwords to authenticate the user on devices for secure communication. Speech processing technology in voice identification may ensures and secure the user access over a range of systems, devices, and services. The formant frequency is the resonant frequency of the vocal cord. Frame-to-frame formants achieved using linear predictive coding (LPC) analysis technique by obtaining the tops of the envelopes. The source voice for vowel is a periodic signal in combination of fundamental frequency and a random noise generator generates unvoiced consonants. In this study, we used voice test samples of different male speakers from age 15 to 20 years. From each speaker utterance of different Indian English words, including vowels and consonants recorded using a digital audio editor software GoldWave v6.57. Formant frequency extracted from the spectrogram of recorded words using MATLAB 2016a signal processing toolbox. Finally, presented an approach for extracting vowels in the words spoken based on three formant frequencies such F1, F2 and F3. The results has shown the significance of vowels in Indian English words when formant frequency of vocal tract is considered.

Keywords: Indian English, vowels, formants, speech recognition, vocal tract, LPC.

1. Introduction

Speech is something seen through the movement of the organs of speech - phonetics of speech, or by listening to real sounds - the phonetics of hearing. Speech production consists mainly three functions such as motor control, articulatory Motion and sound generation. In terms of speech processing, acoustic wave produced by the mouth and nose reaches the human ear and sensed by sensory nerves that connect the human ear and brain [1-2]. The source voice for vowel is a periodic signal in combination of fundamental frequency and a random noise generator generates unvoiced consonants. Formant synthesis involves speech using specific tools governed by a set of rules. It does not involve any recorded data for signals but based on other parameters such as frequency, amplitude etc. [3]. The frequency, bandwidth, and amplitudes can define formants. Formant amplitude can be analysed by format frequency, bandwidth, and active source power [4]. Formant areas in the speech signal usually measured by making a computer LPC. The highest altitudes measured by root or computer-based solid spectrum and by size [5]. Duration of the speech in transfer of linguistic information considered as an important factor. The length of vowels remains between 50ms to 100ms in case of preceding voiced consonants and for consonants up to 70ms. To improve speech conversion for deaf, auto-translation, and defence, speech recognition systems need improvements [6-7].

Journal of Physics: Conference Series **2236** (2022) 012011 doi:10.1088/1742-6596/2236/1/012011

1.1. Signal Processing
The most common approach of signal processing is the frequency Domain. Sound wave is period signal, subjected to analysis of Fourier Transformation, which is conversion of signal from time domain to the frequency domain g(t) = G(f). For a speech signal, the speaker is the information source. The measurement records in terms of acoustic waveform.

1.2. Vocal System and vowels
Speech constructed with three levels linguistics, acoustics and physiology. At the level of linguistic brain considered, at acoustic level the sound waves and at physiological level vocal muscular. The vocal system includes human vocal organs such as Nasal cavity, Hard palate, Alveolar ridge, Soft palate, Apex, Dorsum, Uvula, Radix, Pharynx, Epiglottis, False vocal cords, Vocal cords, Larynx, Oesophagus, and Trachea. An example of speech production for human mentioned in Figure 1.

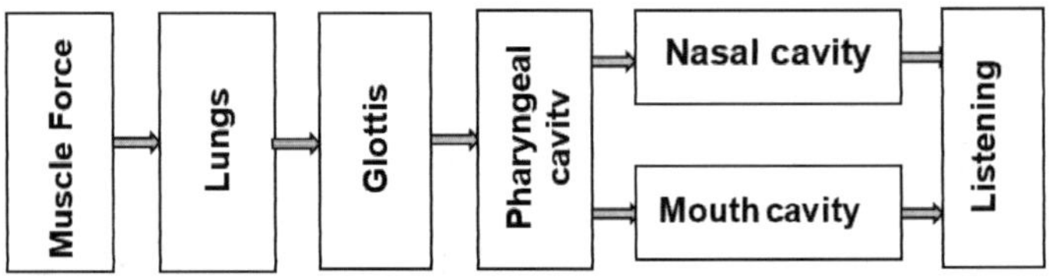

Figure 1. Speech production system for human.

Vowels include the production of sounds with no obstruction through the vocal tract. Three main organs such as lips, lower jaw and the tongue are the primary criterion in vowels. The example of vowels includes 'a', 'e', 'i', 'o' and 'u' used in Indian English speaking.

1.3. Formant frequency
The formant frequency is the resonant frequency of the vocal cord. Vowels sounds produced acoustically filtered through quasi-periodic air pulses as they spread through the vocal cords. These format frequencies help to distinguish phonemes when acoustic conditions are different. Vowels correspond to the most powerful regions of the spectrum and identified by the peaks of the spectral envelope [8]. First and second formants such as F1, F2 frequency of vowel status considered important signs of vowel recognition. Format frequencies, classification of F1 and F2, are the main features of vowel identification [9]. Figure 2 shows the formants in a sound sample such as F1, F2 and F3.

1.4. Speaker Identity and characterization
Speeches bear messages through the languages spoken by the speakers. A speaker identified through the speech or voice spoken. In this contest digital signal processing (DSP) technique play very important role for studying speech signal of an individual. In speaker identification, the prime interest is not recognizing the words but determining who is speaking the words. Speaker Characterization involves the tasks, which consists in using a speech signal to extract important features involved in the identity of the speaker who utters it.

1.4.1 Speaker identity applications
- System passwords or codes security.
- Identification of the speaker in a noisy room by matching different formats.
- Speaker identification systems have potential forensic applications.
- Finding further possibility of research in the field of speech technology.
- Possibilities of better understanding of different languages.

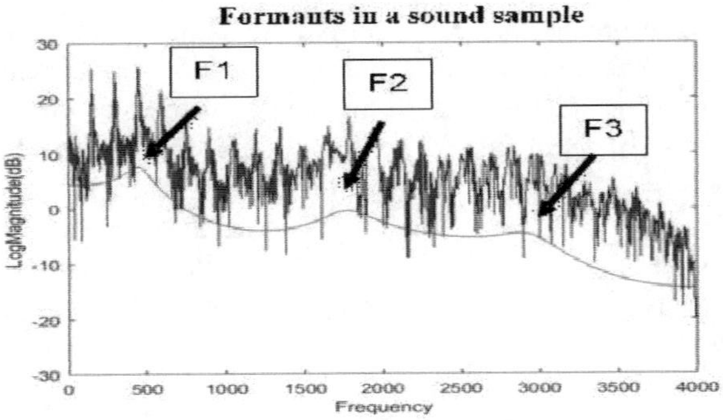

Figure 2. Example of Formants in a sound sample.

1.5. Objective of Research
Features of speech signal indicate the speaker's identity. The speaker identification techniques may be useful to identify a speaker over the internet applications such as voice over internet protocols. Further, may lead to develop software applications and hardware design at chip level specifically useful to secure voice-based systems. This research work is carried out to define speech based new features & establishment of techniques.

1.6. Limitations of speech processing research
With the growth of mobile technology, a growing body of research confirms that advanced voice authentication systems are at risk of replay attack, when the intruders deceive voice verification systems with pre-recorded or integrated voice samples by a real user [10]. The more number of analysis remains outside the scope of present methodologically based study. Further, because of strong performance of logistic regression neural networks found out of scope [11]. The advantages of magnitude analysis over format analysis are most evident in automated speech recognition systems. The great disadvantage of this application is that the interaction between sizes is ignored [12]. By using the hearing response independently, it may be impossible to determine the actual size of the lips required to sound 'u' [13]. The status has reflected downfall decrease in speech capacity and rise in number of unvoiced components in comparison with neutral state [14]. The results not improved with the use of MFCC. The hamming window used for speech processing because its spectrum falls off quickly [15]. The robotic collects data about jaw movement to produce headwords, but jaw movements can change vowel structures [16].

This paper is organized in five sections. After discussion with introduction part, second section describe the related works. The research methodology mentioned in section three. In the section, four results and discussion presented. Finally, the paper concluded with section five.

2. Related Works
Human ability to distinguish sound intensity reflected in their production mechanism. The American English vowel was analysed in terms of production and perception. The authors have found participants in a group of high-vision skills to produce a vowel with a greater degree of comparison [17]. Some tend to support theoretical predictions and evaluating formant frequency in human identity. Different types of acoustical information in baboon grunts and humans shown continuity with intensity of vocal affect [18]. Reliable format frequency measurements is a very challenging task to evaluate from the speech signal. Different types of experiments have shown the probability of a linear relationship among formant and spectral evaluations. Key Component Analysis used to find the most descriptive features of the vowel variation defined by log energy. Authors observed that the formant bandwidth can be predicted using the same method [19]. The authors attempted to make the vocabulary recognition with the use of formant analysis in the Serbian language and found out about

vowels spoken. The standard way of distinguishing vowels explained by utilizing formant frequencies [20]. The relationship between breathing and voice communication analysed in measurement of formants reflecting formation of voice signal. The authors' study included smoking habits of individuals according to formant frequencies. Audio recordings are analysed using the software Praat for extraction of formants [21]. In the study, the authors performed two experiments with the production of a dynamic frequency and audible frequency. The authors observed F1, F2, age and system response with internal connection [22]. The sizable detection model based on stabilized wavelet transform (SWT) to define the different sizes of discriminatory data was proposed. The point of subjective quality (PSE), shown a variation in imagined value, changed about 10% of the spoken speech sounds [23].

3. Methodology

3.1. Linear Predictive Coding (LPC)

Researchers in speech processing use the LPC base Formant measure technique. In this method, all ranges of frequencies partitioned in the fixed number of segments and every component shows the frequency of the formant. The order of every partition 'k' defined with boundary. A polynomial called Fourier transform provided as predictor [24]. LPC is a way to make the signal source model stand out in speech signal processing and to have extensive use in different fields of speech analysis.

LPC known as a fast way to analyse voice. The glottis is an opening in vocal folds in the larynx. The throat, mouth, and nose (vocal cords) represent a tube with a range of resonance frequencies known as formants. A difference number to identify signal samples and linear predictor known as Linear Predictive Coding [25]. The difference number coefficients (predictive coefficients) reflect formants. In case of LPC model, the speech sample represent s(n) during n.

$$s(n) \approx a_i s(n-1) + \cdots + a_n s(n-p) \quad (1)$$

In the equation (1) a_1, a_2, a_n as the coefficients remain constants in speech analysis. The equation (1) represented as below,

$$\text{s(n)} = Gu(n) + \sum_{i=1}^{p} a_i s(n-1) \quad (2)$$

Where Gu(n) is excitation term, *u(n)* excitation normalization and G excitation gain.

3.2. The LPC analysis

The relationship in s(n) & u(n) defined based on Figure I.

$$\text{s(n)} = Gu(n) + \sum_{i=1}^{p} a_k s(n-1) \quad (3)$$

By considering the samples of a previous speech as the s(n) standard defined as,

$$\hat{s} = \sum_{k=1}^{p} a_k s(n-k) \quad (4)$$

The predictor error $e(n) = s(n) - \hat{s}(n)$ (5)

The transmission function $X(Y) = 1 - \sum_{k=1}^{p} a_k Y^{-k}$ (6)

The problems with LPC analysis is to specify a predictor coefficient 'a_k' in order to make digital filter speech features similar to speech wave format in analysis window.

3.3. Operating system

Operating System: Windows 10 Pro, Manufacturer: Dell, Model: Vostro 470, Processor: Intel® Core™ i7-3770 CPU @ 3.40GHz, installed memory (RAM): 8.00 GB, System Type: 64-bit Operating System.

3.4. Voice recording

HP Headphone and Microphone, 100-16Khz, Headphone 20-20Khz, Impedance 32 Ohm, Power Input 50mW Sensitivity 108DB. In this research work, the voice sample recorded as .wav formats at sampling rates of 8 Hz because frequency up to 8 KHz can be represented (Nyquist Theorem). The data format. remained 16-bit signed integers as it is common for .wav files Total ten (10) number of words are considered for voice recording from 7(seven) number of speakers, accordingly, 70 number of sample were collected from the male speakers between 15 to 20 year of age. Table 1 presents the words used for voice recording for vowels.

Table 1. Type of words used for voice recording for vowels

Pronouncement	Vowels	Pronouncement	Vowels
aeroplane	'O'	India	'I'
bulb	'U'	mobile	'E'
car	'A'	rocket	'O'
computer	'O'	table	'A'
fan	'A'	train	'I'

3.5. GoldWave software.

Goldwave v6.57 is a professional full featured digital audio editor used to play, record, import, edit, restore, process, analyse and convert audio on the computer system. The GoldWave compose of three windows: The main window, sound window, and control window. This software used for recording the voice samples at required specifications. The Goldware software sound recording window shown in Figure 3.

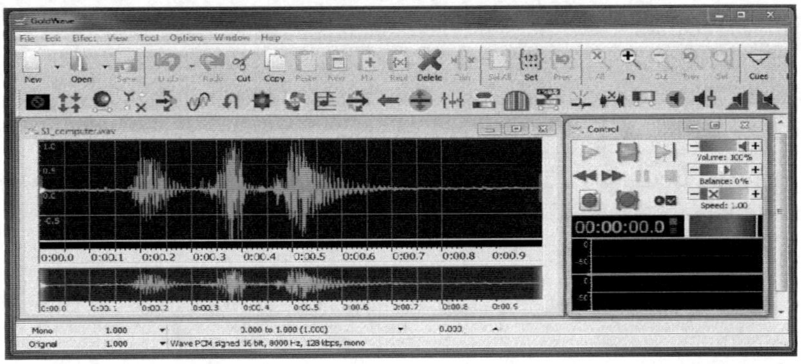

Figure 3. Goldwave software sound recording window

3.6. MATLAB

Matlab is a powerful computational environment as s/w package with advanced built-in features and a standard tool for engineering disciplines. MATLAB available with many signal processing toolbox, for speech signal processing, the signal processing tools box is most useful. MATLAB 2016a used for formant analysis of recorded voice samples.

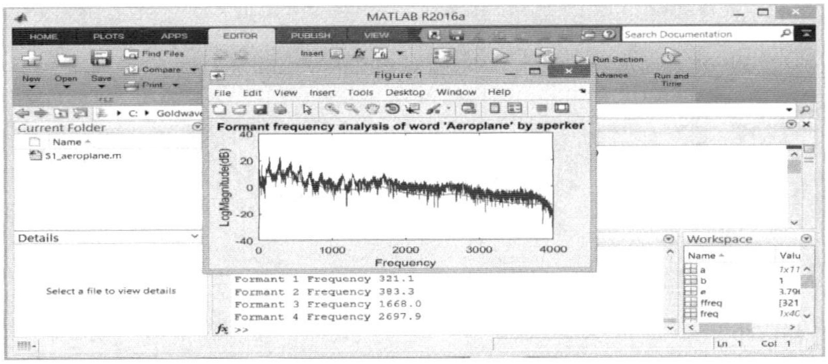

Figure 4. Formant analysis of sound samples

Different methods are applicable in speech processing. In this research the sound samples in the form of different words are analysed for extraction of formants from the vowel 'a','e','i','o' and 'u' with liner predictive method (LPC) method using MATLAB to present different values of speech data. Figure 4 shows the formant analysis of sound samples.

4. Results and discussion

This research work focus on the features found in the speech signal such as formants F1, F2 and F3 to determine the identity of the speaker from recorded vowel based voice samples (for example 'a' in aeroplane, 'e' in rocket, 'i' in mobile, 'o' in rocket etc.). In this work ten number of words spoken by seven number of male speakers processed using voice signal processing techniques. Total 70 number of samples analysed for three formants. The initial formants F1, F2 and F3 shown in Figure 5 to Figure 10 for speaker 1, speaker2, speaker3, speaker4, speaker5 and speaker 6 respectively.

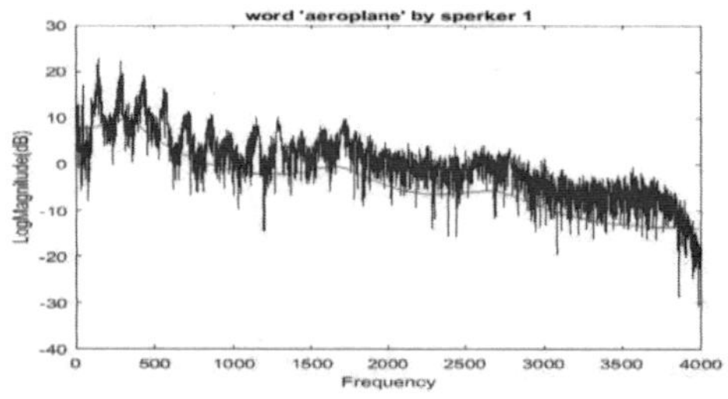

Figure 5. Formant frequency of the word 'aeroplane' by speaker 1.

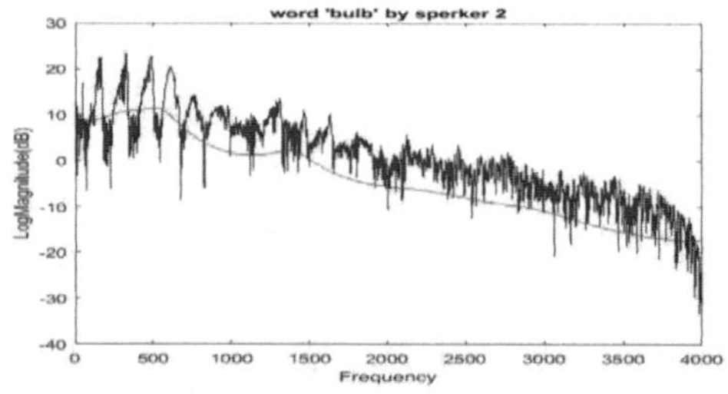

Figure 6. Formant frequency of the word 'bulb' by speaker 2.

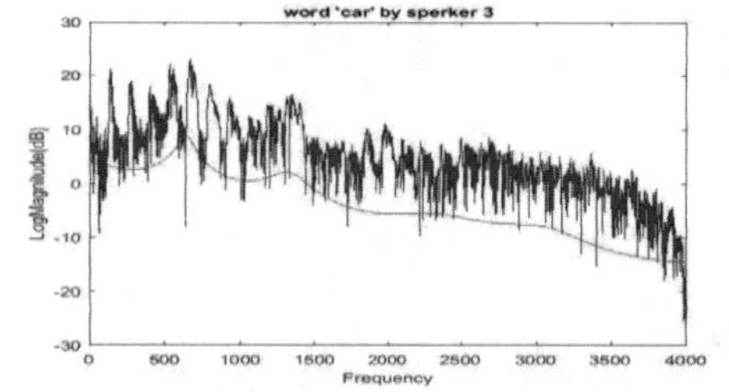

Figure 7. Formant frequency of the word 'car' by speaker 3.

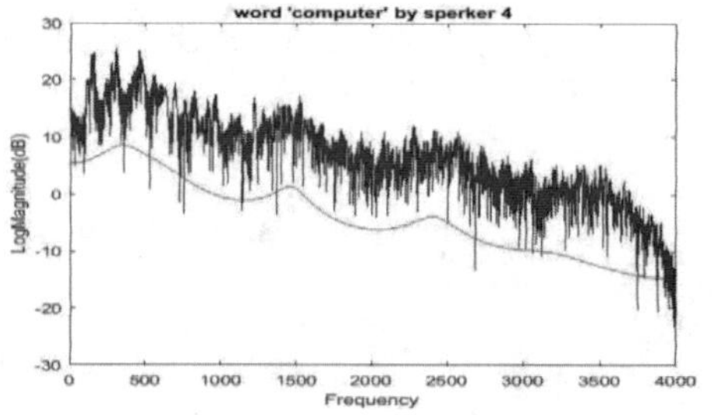

Figure 8. Formant frequency of the word 'computer' by speaker 4.

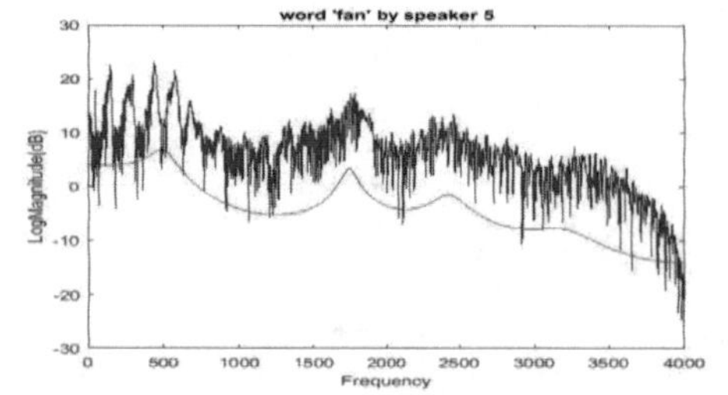

Figure 9. Formant frequency of the word 'fan' by speaker 5.

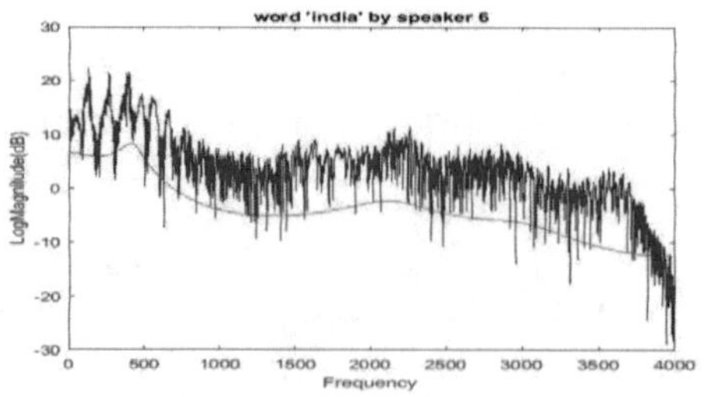

Figure 10. Formant frequency of the word 'india' by speaker 6.

Journal of Physics: Conference Series **2236** (2022) 012011 doi:10.1088/1742-6596/2236/1/012011

These formants frequencies recorded and presented in the form of table. The Table 2 shows the three-formant frequencies F1, F2 and F3 for ten number of words spoken by all seven speakers.

Table 2. Data for formant frequency of sound records.

Samples	Formant frequencies (Hz)			Samples	Formant frequencies (Hz)		
	F1	F2	F3		F1	F2	F3
S1 aeroplane	0383.3	1668.0	2697.9	S1 India	0336.8	1686.5	2580.5
S2 aeroplane	0218.6	0606.8	1721.5	S2 India	0496.4	1726.3	2844.6
S3 aeroplane	0535.0	1677.3	2660.6	S3 India	0425.1	2001.3	3169.9
S4 aeroplane	0426.4	1347.4	2882.4	S4 India	0339.4	1600.0	2676.5
S5 aeroplane	0540.0	1515.8	2259.2	S5 India	0551.1	1749.2	2798.2
S6 aeroplane	0521.5	1669.9	2877.8	S6 India	0421.3	2142.1	3016.4
S7 aeroplane	0528.8	1488.7	2398.0	S7 India	0344.7	0175.5	2571.7
S1 bulb	0381.5	1184.6	2739.4	S1 mobile	0314.4	0785.1	1721.7
S2 bulb	0553.8	1389.7	2254.0	S2 mobile	0620.4	1679.1	2727.4
S3 bulb	0577.0	1308.3	2262.4	S3 mobile	0670.3	1459.5	2326.0
S4 bulb	0312.5	0772.5	1825.2	S4 mobile	0657.0	1681.6	3027.3
S5 bulb	0543.7	1326.4	2428.4	S5 mobile	0637.2	1488.5	2408.2
S6 bulb	0267.6	0640.6	1428.6	S6 mobile	0664.1	1614.4	2420.4
S7 bulb	0530.3	1247.1	2340.8	S7 mobile	0293.0	0638.7	1375.8
S1 car	0642.4	1269.3	2509.8	S1 rocket	0337.5	1653.4	2113.2
S2 car	0733.7	1368.9	2524.1	S2 rocket	0462.9	1863.9	2727.2
S3 car	0638.8	1331.4	2330.3	S3 rocket	0524.8	1351.2	3154.6
S4 car	0614.0	1258.5	2649.1	S4 rocket	0405.5	1845.2	2748.5
S5 car	0758.7	1275.2	2457.1	S5 rocket	0462.7	1948.1	2936.8
S6 car	0681.8	1307.1	2308.1	S6 rocket	0472.7	2001.2	2947.7
S7 car	0714.2	1232.6	2333.8	S7 rocket	0358.8	1605.6	2182.0
S1 computer	0322.2	1639.7	2449.7	S1 table	0398.3	1648.4	3117.7
S2 computer	0310.3	1550.5	2641.1	S2 table	0391.4	1308.4	2025.6
S3 computer	0638.8	1331.4	2330.3	S3 table	0520.1	1934.7	2581.7
S4 computer	0614.6	1258.5	2649.1	S4 table	0401.4	1417.3	2411.5
S5 computer	0758.7	1275.2	2457.1	S5 table	0397.7	1387.2	2286.1
S6 computer	0681.8	1307.1	2308.1	S6 table	0526.0	1529.4	2083.2
S7 computer	0714.2	1232.6	2333.8	S7 table	0356.9	1448.9	2336.0
S1 fan	0253.4	0634.6	1712.9	S1 train	0291.8	1649.6	2290.6
S2 fan	0275.1	0620.4	1778.0	S2 train	0242.8	0604.3	1776.4
S3 fan	0589.1	1871.0	2578.7	S3 train	0454.2	1775.3	2938.9
S4 fan	0263.1	0619.7	1710.1	S4 train	0379.0	1586.5	2460.7
S5 fan	0502.4	1741.1	2433.7	S5 train	0555.7	1669.6	2354.4
S6 fan	0662.5	1759.3	2674.7	S6 train	0485.3	1936.1	3092.4
S7 fan	0269.8	0622.0	1584.0	S7 train	0322.3	1687.1	2407.8

From the Table 2 the graphs plotted for analysis of speech features. The Figures from Figure 11, to Figure 18 shows the comparison of three formants frequencies of words 'aeroplane', 'bulb', 'fan', 'india', 'mobile', 'rocket' and 'train' respectively by all seven speakers.

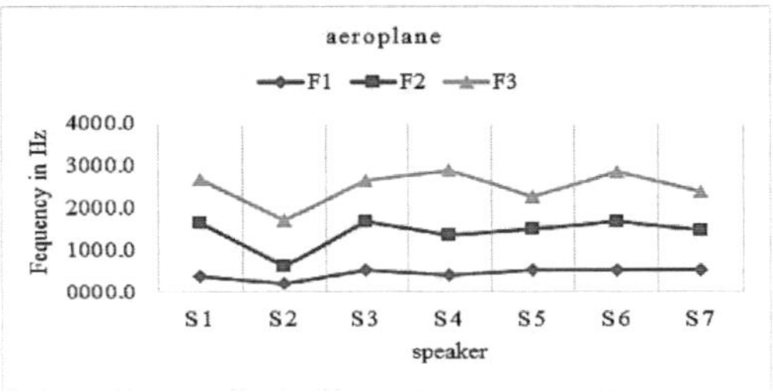

Figure 11. Formant comparison of word 'aeroplane' by seven speakers.

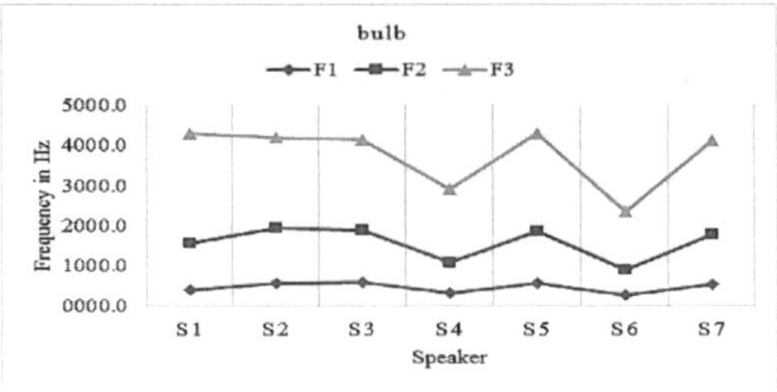

Figure 12. Formant comparison of word 'bulb' by seven speakers.

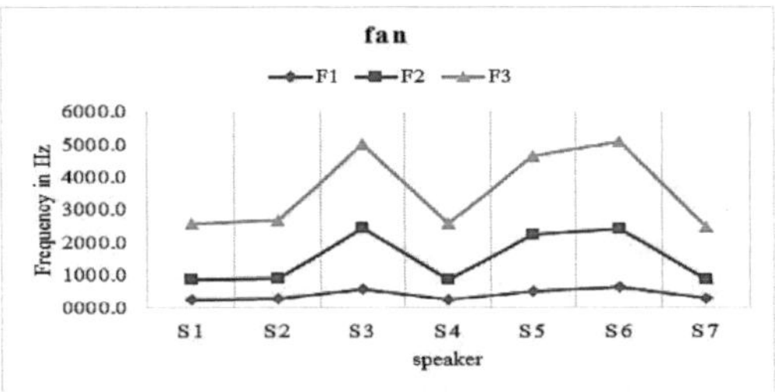

Figure 13. Formant comparison of word 'fan' by seven speakers.

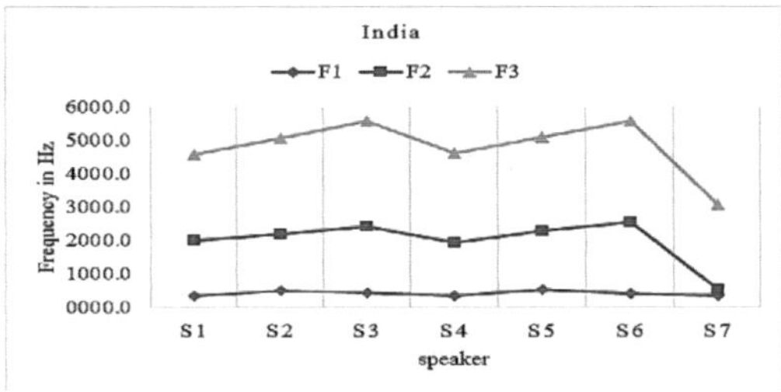

Figure 14. Formant comparison of word 'india' by seven speakers.

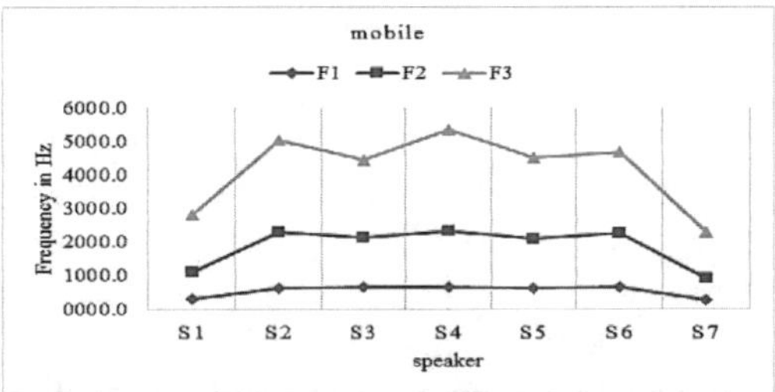

Figure 15. Formant comparison of word 'mobile' by seven speakers.

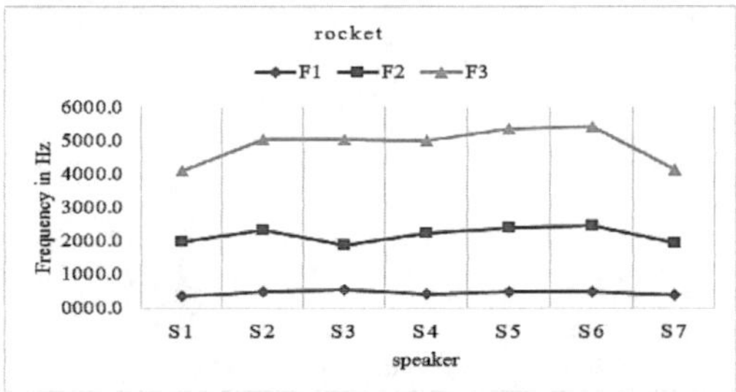

Figure 16. Formant comparison of word 'rocket' by seven speakers.

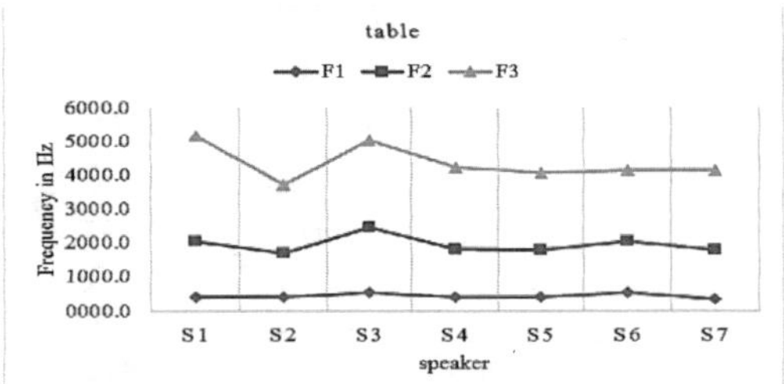

Figure 17. Formant comparison of word 'table' by seven speakers.

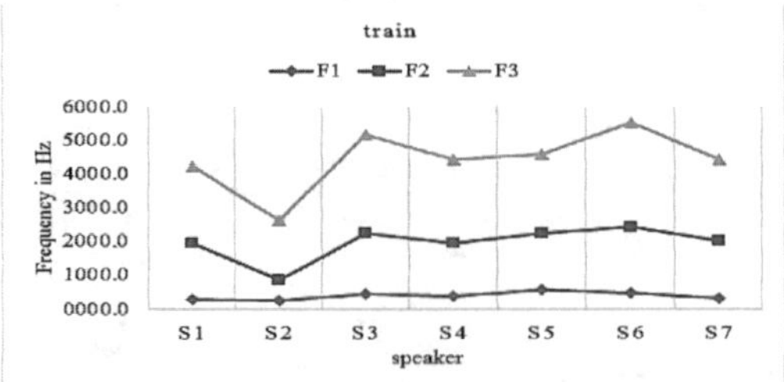

Figure 18. Formant comparison of word 'train' by seven speakers.

It has been observed from the from Figure 11 to Figure 18 that The first formant frequency (F1) is found always higher in case of speaker 3 w.r.t rest 6(six) speakers for Indian English word utterances. The effect seems more pronounced in terms of 2nd formant frequency (F2). From the present study, it is found that the magnitude of shifting of F1 to F2 and F2 to F3 much higher in case of speaker 3 as compared to other speakers. The Present study depicts that distinctive feature of speaker's utterances of English words are dependent on the vowel present in the Indian English words.

The Present study reveals that vowel 'a' seems to play typical role in the Indian speaker's utterances. A gradual increase in F1 & F2 for some typical words (in English) also found as a common characteristic feature of speaker 2 and speaker 4. Of course, this effect observed in 60% to 70% cases in the present study. The third formant frequency (F3) found very prominent, in case of speaker 1 speaker 2 and speaker 4.

5. Conclusion

Speech processing technology in voice identification may ensures and secure the user access over a range of systems, devices, and services. The source voice for vowel is a periodic signal in combination of fundamental frequency, which reflects the shape of vocal tract. Frame-to-frame formats achieved using MATLAB analysis. The three formants of each voice sample with ten number word presented in the form of table. Further, the formants compared with all the speakers. The present spectral study on the voice characteristics of male speakers observed that 1st formant and 2nd formant frequency (F1&F2) bears distinctive features in the vocal field and vocal tract. These two parameter could be used to a measurable extend as role parameters for speaker recognition considered in this research work. The location of vowels in words seems playing a significant role in case of male speaker's age from 15 to 20 years. Of course, further in-depth study is required for effective conclusion. To establish this effect more realistically, further study through large volume of common words utterances should be required.

Acknowledgement

This research work not supported by any public or private organization. Authors wish to thank Department of Computer Science and Engineering, NERIST, Arunachal Pradesh for providing lab facilities used in this study.

References

[1] Tatham M, and Morton K 2006 *Speech production and perception*. New York: Palgrave Macmillan
[2] Mahendru H C 2014 Quick review of human speech production mechanism *Int. Journal of Engineering Research and Development* **9(10)** pp 48-54
[3] Chala S 2021 *Text to Speech Synthesizer for Afaan Oromo Using Deep Neural Network* (Doctoral dissertation, ASTU)
[4] Kent R D and Vorperian H K 2018 Static measurements of vowel formant frequencies and bandwidths A review. *Journal of communication disorders* **74** pp 74-97
[5] Vallabha G K and Tuller B 2002 Systematic errors in the formant analysis of steady-state vowels *Speech communication* **38(1-2)** pp 141-160
[6] Koffi E 2021 A Comprehenisve Review of the Acoustic Correlate of Duration and Its Linguistic Implications *Linguistic Portfolios* **10(1)** p 2
[7] Mallie D 2014 *Voice processing using MATLAB as a tool*
[8] Biswas A S T. I K Sahu P K, Bhowmick A and Chandra M 2014 Hindi vowel classification using GFCC and formant analysis in sensor mismatch condition *WSEAS Trans Syst*, **13** pp 130-143

[9] Jin S H and Liu C 2013 The vowel inherent spectral change of English vowels spoken by native and non-native speakers *The Journal of the Acoustical Society of America* **133(5)** pp 363-369

[10] Zhang L 2021 *Towards Enhanced Mobile Voice Authentication* (Doctoral dissertation, The Florida State University)

[11] Ortiz A J, Toles L E, Marks K L, Capobianco S, Mehta D D, Hillman R E and VanStan J H 2019 Automatic speech and singing classification in ambulatory recordings for normal and disordered voices *The Journal of the Acoustical Society of America* **146(1)** pp 22-27

[12] Pols L C, Tromp H R and Plomp R 1973 Frequency analysis of Dutch vowels from 50 male speakers *The journal of the Acoustical Society of America* **53(4)** pp 1093-1101

[13] Murakami M, Kröger B Birkholz P and Triesch J 2015 Seeing [u] aids vocal learning: Babbling and imitation of vowels using a 3D vocal tract model, reinforcement learning, and reservoir computing *IEEE Int. Conf. on Development and Learning and Epigenetic Robotics (ICDL-EpiRob)* pp. 208-213

[14] Barreda-Ángeles M, Aleix-Guillaume S and Pereda-Baños A 2020 Users' psychophysiological, vocal, and self-reported responses to the apparent attitude of a virtualaudience in stereoscopic 360-video *Virtual Reality* **24(2)** pp 289-302

[15] Abdulsatar A A, Davydov V V, Yushkova V V, Glinushkin A P and Rud, V Y 2019 Age and gender recognition from speech signals *Journal of Physics: Conference Series IOP Publishing* **1410 (1)** p. 012073

[16] Scott S K 2021 The neural control of volitional vocal production—from speech to identity, from social meaning to song. *Philosophical Transactions of the Royal Society* **377(1841)** p 20200395

[17] Cheng H S, Niziolek C A., Buchwald A and McAllister T 2021 Examining the relationship between speech perception, production distinctness, and production variability *Frontiers in Human Neuroscience* **15** pp 1-12

[18] Rendall D 2003 Acoustic correlates of caller identity and affect intensity in the vowel like grunt vocalizations of baboons *The Journal of the Acoustical Society of America* **113(6)** pp 3390-3402

[19] Hogberg J 1997 Prediction of formant frequencies from linear combinations of filter bank and cepstral coefficients. KTH-STL Quarterly Progress *Rep Royal Inst. Technol. Stockholm* Sweden pp 41-49

[20] Prica B and Ilić S 2010 Recognition of vowels in continuous speech by using formants Facta universitatis-series *Electronics and Energetics* **23(3)** pp 379-393

[21] Suwandi G R F, Mustajab M A, Haekal M, Khotimah S N and Haryanto F 2020 An Acoustic Analysis of Formants between Frequently Smoking Subjects and Non-smoking Subjects *Journal of Physics: Conf. Series IOP Publishing* **1505(1)** p 012053

[22] Ilyas M and Naït-Ali A 2019 Machine Learning Based Detection of Hearing Loss Using Auditory Perception Responses *IEEE 15th Int. Conf. on Signal-Image Technology & Internet-Based Systems (SITIS)* pp 146-150

[23] Matsui T, Irino T, Uemura R, Yamamoto K, Kawahara H and Patterson R D. 2022 Modelling speaker-size discrimination with voiced and unvoiced speech sounds based on the effect of spectral lift *Speech Communication* **136** pp 23-41

[24] Singha M K, Boro J Sarma B 2012 LPC analysis of vowels and formant analysis of some typical CV and VC type of words in Bodo language *IEEE 3rd National Conf. on Emerging Trends and Applications in Computer Science* pp 193-197

[25] Shahriar S and Hoq M N 2016 Evaluation of LPC trajectory for Vowel-Consonant-Vowel sequence *IEEE 19th Int. conf. on Computer and Information Technology (ICCIT))* pp 235-239

Journal of Physics: Conference Series **2236** (2022) 012012 doi:10.1088/1742-6596/2236/1/012012

A Compact Wide Stopband Band Pass SIW Filter For K-Band Applications

M Edison Singh[1]

[1]ECE Department, North Eastern Regional Institute of Science and Technology, Nirjuli, India -791109

em@nerist.ac.in

Abstract. Direct coupled filters have narrow stopband due to higher order modes passband. A compact dual layer band pass filter is proposed with wide stop band. The proposed filter provides the return loss of 16.87db and insertion loss of 2.5dB at the passband frequency 20.9 GHz. The filter size is half of the conventional direct coupled filters. The stop band covers up to 40 GHz with insertion loss below 35dB. The filter has no slot which makes it immune to the EMI/EMC problems.

1. Introduction

Filters play an important part of the transceiver system. As the technology advances, the demand for filter with compact size, wide stop band, high selectivity, low cost etc. has been increased. Since the introduction of Substrate Integrated Waveguide (SIW), it has become popular among the filter designer as it offers the advantage of a planar structure like micro strip line but relatively high-quality factor similar to the waveguide. But SIW resonator based direct coupled filter suffers from low selectivity and narrow stop band due to the second pass band which arises at the next higher order mode. The works [1][2] on the paper used slots to radiate the unwanted higher order mode. This will cause EMI/EMC problems. In paper [3], oversize cavity is used to create multiple coupling path to create transmission zero to improve selectivity. But the next higher order mode is not suppressed. The works on paper [4] is based on the weak coupling of modified higher TM mode of the circular cavity. Substrate integrated folded waveguide is excited with strip line feed with odd symmetric mode to suppress mode with even symmetry and inset to tune out higher order modes [5]. [6] uses mode dependent coupling aperture between two cavities stack vertically to suppress unwanted mode. In this paper, a compact second order band pass filter with wide stop band is proposed. In order to achieve the wide stop band, it is shown that the second pass band is suppressed. This is achieved by weakening the coupling between source/load and resonators and the coupling between the resonators for the higher order modes. To achieve the compactness, it is implemented in two layers.

2. Design and Results

Here a second order direct coupled filter is considered. The general coupling matrix of the filter is shown in figure 1 where M represent the frequency independent coupling. The second order filter designed using SIW square cavity shown in figure 2.

$$\begin{bmatrix} 0 & M_{S1} & 0 & 0 \\ M_{S1} & 0 & M_{12} & 0 \\ 0 & M_{12} & 0 & M_{2L} \\ 0 & 0 & M_{2L} & 0 \end{bmatrix}$$

Figure 1. General Coupling Matrix of the filter.

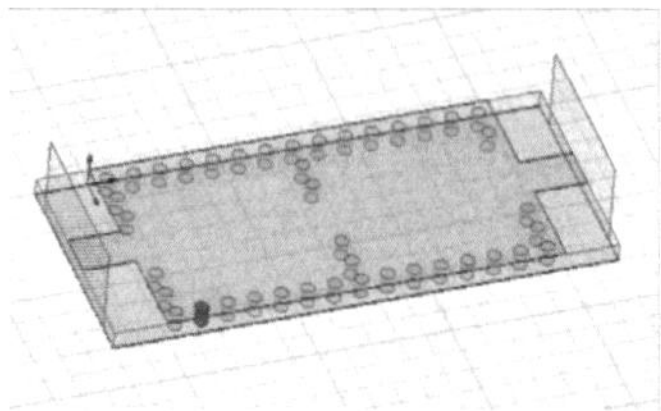

Figure 2. Second order direct coupled filter.

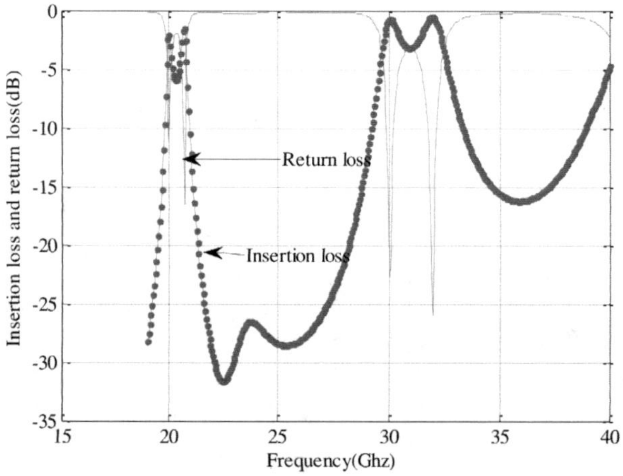

Figure 3. The response of the second order direct coupled filter.

The response of the filter is shown in figure 3. The fundamental passband is observed at 20.9GHz due to TE_{10} mode and the second passband is observed 30GHz due to TE_{12} mode. This second pass band reduces the stopband bandwidth. The second passband arises because the resonators are coupled though higher order modes at 30GHz. The higher order modes are coupled since side coupling of the resonator is used. When side coupling is used, the two resonators are coupled due to the nature of the electric field distribution of square cavity resonator shown in figure 4 and figure 5.

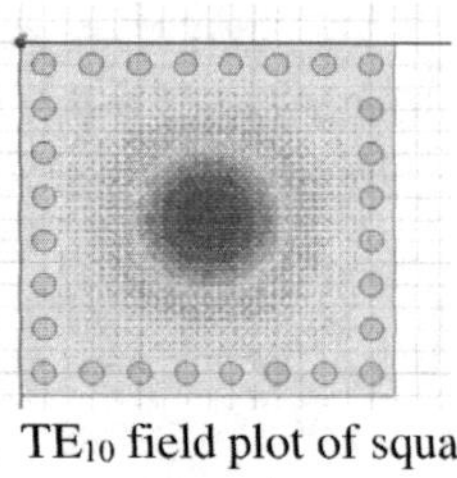

Figure 4. TE_{10} field plot of square cavity resonator at 20.9 GHz .

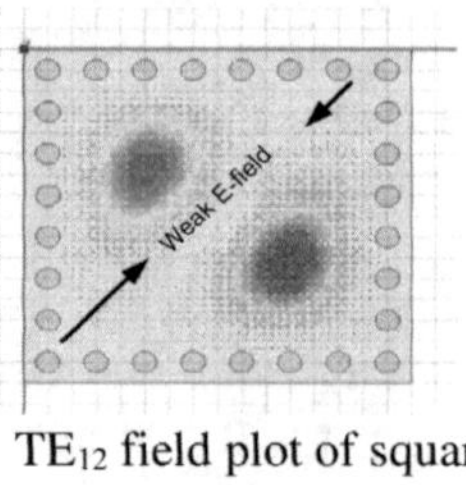

Figure 5. TE_{12} field plot of square cavity resonator at 30GHz.

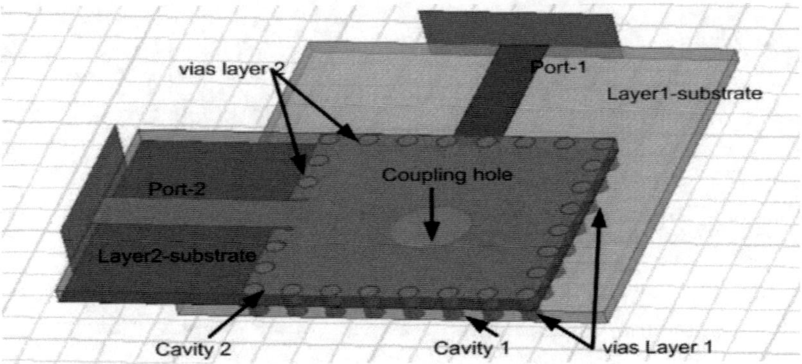

Figure 6. The proposed filter.

To improve the performance of the filter, the pass band arises from the next higher order mode must be removed. To remove the second pass band due to the next higher mode, the coupling coefficient for the fundamental mode of the filter has to be kept remains unaffected but the coefficient for the second mode between two resonators or source/load and resonator needs to be minimized. It can be seen that the fundamental mode TE_{10} is symmetric around the center while the next mode TE_{12} has null electric filed along the center in figure 4 and figure 5. So, two resonators coupled through a circular slot at the center will reduce the TE_{12} mode coupling while the TE_{10} mode coupling is strong. Further the second port is placed in such a way that it lies in the null E field region of the TE_{12} mode. The proposed second order filter is shown in figure 6. The substrate used is Rogers RT/Duroid 5880(tm). The size of the square cavity is 8mmX8mm. The two resonators form two layers. The resonators are coupled via a circular hole (radius 1mm) in the ground of the top resonator. Also, the second port is placed at the weak field region of the second mode. The simulated response of the proposed filter in HFSS is shown in figure 7. From the S_{21} plot it can be seen that the second passband has been suppressed below 35 dB improving the stopband of the filter. The field plot of the filter at the fundamental mode and the higher order mode frequencies are shown in figure 8 and figure 9. In the passband the S_{11} is -16.87db and S21 is -2.5dB. The band stop reaches up to 40GHz considering below -35dB. A comparison table with the previous work is given in table 1.

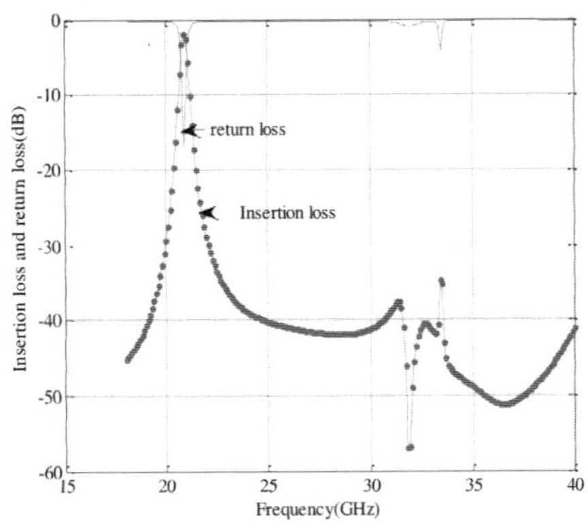

Figure 7. Insertion loss and return loss plot of the proposed filter.

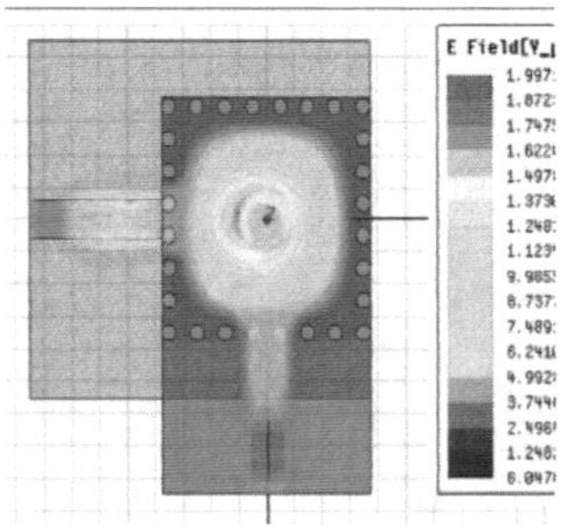

Figure 8. Field plot of the proposed filter at 20.9GHz .

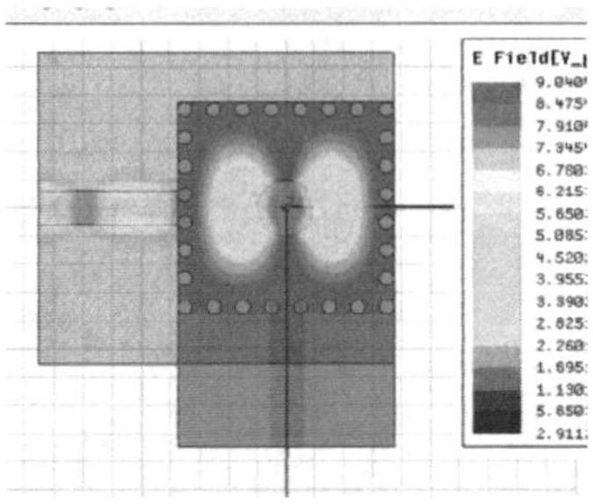

Figure 9. Field plot of the proposed filter at 30 GHz.

Table 1. Comparison table.

	Centre frequency	Order	Stopband / Insertion loss
[1]	9.7GHz	5	Up to 16GHz /-20dB
[2]	10GHz	3	Up to 19 GHz /-35dB
[3]	20GHz	4	Up to 31 GHz/-20dB
[4]	22GHz	2	Up to 26 GHz/-20dB
This Work	20.9GHz	2	Up to 40 GHz /-35dB

3. Conclusion

A compact second order SIW filter is designed for K-band applications having a passband at 20.9GHz with wide stop band up to 40 GHz with insertion loss below 35dB. The filter is implemented using Rogers RT/Duroid 5880(tm) as substrate. Using 2 layers, the lateral area of the filter is reduced to half of the conventional direct coupled filter. In the passband the return is 16.87db and insertion loss is 2.5dB. As no slot are used to suppress the higher order mode passbands, it provides immunity to EMI/ EMC problems.

References

[1] Mehrshahi, E., Rezaiesarlak, R. and Salehi, M. 2011 *Int J RF and Microwave Comp Aid Eng*, 21: 234-239.

[2] Salehi, M. and Mehrshahi, E. 2011 *Int J RF and Microwave Comp Aid Engg*, 21: 650-657

[3] X. Chen, K. Wu and D. Drolet 2009 *IEEE Transactions on Microwave Theory and Techniques*, vol. 57, no. 3, pp. 674-683

[4] B. Lee, T. Lee, K. Lee, M. Uhm and J. Lee 2015 *IEEE Microwave and Wireless Components Letters*, vol. 25, no. 6, pp. 367-369

[5] R. Moro, S. Moscato, M. Bozzi and L. Perregrini *IEEE Microwave and Wireless Components Letters*, vol. 25, no. 4, pp. 214-216

[6] D. Jia, Q. Feng, Q. Xiang and K. Wu 2016 *IEEE Microwave and Wireless Components Letters*, vol. 26, no. 9, pp. 678-680

J. Phys.: Conf. Ser. **2236** (2022) 012013 https://doi.org/10.1088/1742-6596/2236/1/012013

Retraction

Retraction: Deep Learning-Based Clinical Image Segmentation Methods With Different Modalities (*J. Phys.: Conf. Ser.* **2236** 012002)

Published 11 May 2022

This article has been retracted by IOP Publishing following an allegation that this article may contain tortured phrases [1], masking overlap of other work [2].

IOP Publishing has investigated in line with the COPE guidelines, and agree this article should be retracted.

IOP Publishing wishes to credit the Problematic Paper Screener for bringing the issue to our attention.

The authors agree to this retraction.

[1] Cabanac G, Labbe C, Magazinov A, 2021, arXiv:2107.06751v1

[2] Sumithra M, Malathi S, 2018, A Survey on Medical Image Segmentation Methods with Different Modalitites, IJERT. **6** 2

Retraction published: 11 May 2022

AUTHOR INDEX